# Glencoe
# Secondary
# Mathematics

## ALIGNED TO THE

COMMON

CORE

STATE

STANDARDS

## Algebra 1

 **Education**

*Bothell, WA • Chicago, IL • Columbus, OH • New York, NY*

TI-Nspire is a trademark of Texas Instruments Incorporated.
Texas Instruments images used with permission.

**connectED.mcgraw-hill.com**

Send all inquiries to:
McGraw-Hill Education
STEM Learning Solutions Center
8787 Orion Place
Columbus, OH 43240

ISBN: 978-0-07-661900-9
MHID: 0-07-661900-1

Printed in the United States of America.

3 4 5 6 7 8 9 QDB 19 18 17 16 15 14 13 12

McGraw-Hill is committed to providing
instructional materials in Science, Technology,
Engineering, and Mathematics (STEM) that give
students a solid foundation, one that prepares
them for college and careers in the 21st Century.

Common Core State Standards
# Table of Contents

# Welcome to Glencoe Secondary Mathematics to the Common Core

## How to Use This Supplement

This supplement is your tour guide to understanding how Glencoe Secondary Mathematics programs teach the new Common Core State Standards. Its purpose is to help make a smooth transition from your state standards to the new Common Core State Standards.

## Crosswalk

The crosswalk is your guide to understanding how to use your current *Glencoe Algebra 1* program with this supplement to create a Common Core State Standards curriculum. Pages vi–xii of the teacher edition show you which lessons in your textbook should be kept, which can be considered optional, which lessons have additional available content, and how the new material fits into the flow of the chapters you already use.

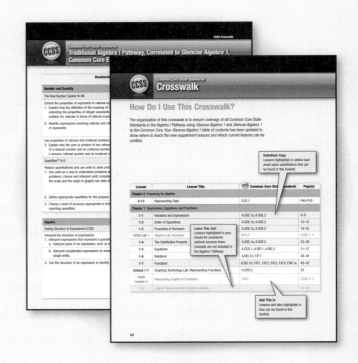

## Correlations

*Glencoe Algebra 1* and *Glencoe Secondary Mathematics to the Common Core* align your curriculum with the Common Core State Standards and the Traditional Algebra I Pathway. You can use pages xii–xviii of the teacher edition to map each standard to the lesson(s) that address each standard.

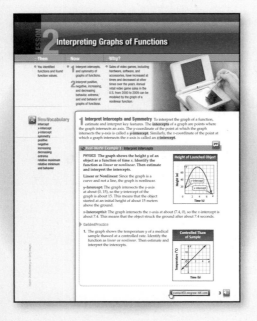

## In This Booklet

*Glencoe Secondary Mathematics to the Common Core* contains additional lessons and labs to address the Common Core State Standards and the Traditional Algebra I Pathway. (See pages 1–78.) You can also find copy for patch substitutions that can help you better meet the Common Core State Standards using your existing program. (See pages 79–83.) Refer to the Crosswalk on pages vi–xii of the teacher edition for appropriate placement of this content in your *Glencoe Algebra 1* textbook.

# Algebra Lab
# Accuracy

All measurements taken in the real world are approximations. The greater the care with which a measurement is taken, the more accurate it will be. **Accuracy** refers to how close a measured value comes to the actual or desired value. For example, a fraction is more accurate than a rounded decimal.

## Activity 1   When Is Close Good Enough?

**Measure the length of your desktop. Record your results in centimeters, in meters, and in millimeters.**

### Analyze the Results

1. Did you round to the nearest whole measure? If so, when?
2. Did you round to the nearest half, tenth, or smaller? If so, when?
3. Which unit of measure was the most appropriate for this task?
4. Which unit of measure was the most accurate?

Deciding where to round a measurement depends on how the measurement will be used. But calculations should not be carried out to greater accuracy than that of the original data.

## Activity 2   Decide Where to Round

a. **Elan has \$13 that he wants to divide among his 6 nephews. When he types 13 ÷ 6 into his calculator, the number that appears is 2.166666667. Where should Elan round?**

   Since Elan is rounding money, the smallest increment is a penny, so round to the hundredths place. This will give him 2.17, and \$2.17 × 6 = \$13.02. Elan will be two pennies short, so round to \$2.16. Since \$2.16 × 6 = \$12.96, Elan can give each of his nephews \$2.16.

b. **Dante's mother brings him a dozen cookies, but before she leaves she eats one and tells Dante he has to share with his two sisters. Dante types 11 ÷ 3 into his calculator and gets 3.666666667. Where should Dante round?**

   After each sibling receives 3 cookies, there are two cookies left. In this case, it is more accurate to convert the decimal portion to a fraction and give each sibling $\frac{2}{3}$ of a cookie.

c. **Eva measures the dimensions of a box as 8.7, 9.52, and 3.16 inches. She multiplies these three numbers to find the measure of the volume. The result shown on her calculator is 261.72384. Where should Eva round?**

   Eva should round to the tenths place, 261.7, because she was only accurate to the tenths place with one of her measures.

## Exercises

5. Jessica wants to divide \$23 six ways. Her calculator shows 3.833333333. Where should she round?

6. Ms. Harris wants to share 2 pizzas among 6 people. Her calculator shows 0.3333333333. Where should she round?

7. The measurements of an aquarium are 12.9, 7.67, and 4.11 inches. The measure of the volume is given by the product 406.65573. Where should the number be rounded?

*(continued on the next page)*

For most real-world measurements, a decision must be made on the level of accuracy needed or desired.

### Activity 3  Find an Appropriate Level of Accuracy

27.5 in.

**a.** Jon needs to buy a shade for the window opening shown, but the shades are only available in whole inch increments. What size shade should he buy?

He should buy the 27-inch shade because it will be enough to cover the glass.

**b.** Tom is buying flea medicine for his dog. The amount of medicine depends on the dog's weight. The medicine is available in packages that vary by 10 dog pounds. How accurate does Tom need to be to buy the correct medicine?

He needs to be accurate to within 10 pounds.

**c.** Tyrone is building a jet engine. How accurate do you think he needs to be with his measurements?

He needs to be very accurate, perhaps to the thousandth of an inch.

### Exercises

**8.** Matt's table is missing a leg. He wants to cut a piece of wood to replace the leg. How accurate do you think he needs to be with his measurements?

**For each situation, determine where the rounding should occur and give the rounded answer.**

**9.** Sam wants to divide $111 seven ways. His calculator shows 15.85714286.

**10.** Kiri wants to share 3 pies among 11 people. Her calculator shows 0.2727272727.

**11.** Evan's calculator gives him the volume of his soccer ball as 137.2582774. Evan measured the radius of the ball to be 3.2 inches.

**For each situation, determine the level of accuracy needed. Explain.**

**12.** You are estimating the length of your school's basketball court. Which unit of measure should you use: 1 foot, 1 inch, or $\frac{1}{16}$ inch?

**13.** You are estimating the height of a small child. Which unit of measure should you use: 1 foot, 1 inch, or $\frac{1}{16}$ inch?

**14.** **TRAVEL** Curt is measuring the driving distance from one city to another. How accurate do you think he needs to be with his measurement?

**15.** **MEDICINE** A nurse is administering medicine to a patient based on his weight. How accurate do you think she needs to be with her measurements?

# Interpreting Graphs of Functions

**::Then**
● You identified functions and found function values.

**::Now**
**1** Interpret intercepts, and symmetry of graphs of functions.

**2** Interpret positive, negative, increasing, and decreasing behavior, extrema, and end behavior of graphs of functions.

**::Why?**
● Sales of video games, including hardware, software, and accessories, have increased at times and decreased at other times over the years. Annual retail video game sales in the U.S. from 2000 to 2009 can be modeled by the graph of a nonlinear function.

 **NewVocabulary**
intercept
*x*-intercept
*y*-intercept
symmetry
positive
negative
increasing
decreasing
extrema
relative maximum
relative minimum
end behavior

**1 Interpret Intercepts and Symmetry** To interpret the graph of a function, estimate and interpret key features. The **intercepts** of a graph are points where the graph intersects an axis. The *y*-coordinate of the point at which the graph intersects the *y*-axis is called a **y-intercept**. Similarly, the *x*-coordinate of the point at which a graph intersects the *x*-axis is called an **x-intercept**.

**PT**

⬤ **Real-World Example 1** **Interpret Intercepts**

**PHYSICS** The graph shows the height *y* of an object as a function of time *x*. Identify the function as *linear* or *nonlinear*. Then estimate and interpret the intercepts.

**Linear or Nonlinear:** Since the graph is a curve and not a line, the graph is nonlinear.

**y-Intercept:** The graph intersects the *y*-axis at about (0, 15), so the *y*-intercept of the graph is about 15. This means that the object started at an initial height of about 15 meters above the ground.

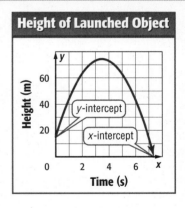

**x-Intercept(s):** The graph intersects the *x*-axis at about (7.4, 0), so the *x*-intercept is about 7.4. This means that the object struck the ground after about 7.4 seconds.

▶ **Guided**Practice

1. The graph shows the temperature *y* of a medical sample thawed at a controlled rate. Identify the function as *linear* or *nonlinear*. Then estimate and interpret the intercepts.

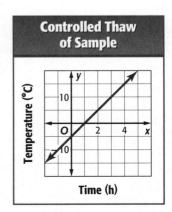

The graphs of some functions exhibit another key feature: symmetry. A graph possesses **line symmetry** in the $y$-axis or some other vertical line if each half of the graph on either side of the line matches exactly.

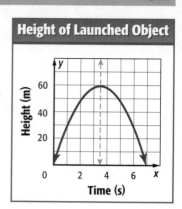

**● Real-World Example 2  Interpret Symmetry**

**PHYSICS  An object is launched. The graph shows the height $y$ of the object as a function of time $x$. Describe and interpret any symmetry.**

The right half of the graph is the mirror image of the left half in approximately the line $x = 3.5$ between approximately $x = 0$ and $x = 7$.

In the context of the situation, the symmetry of the graph tells you that the time it took the object to go up is equal to the time it took to come down.

**Height of Launched Object**

▶ **Guided**Practice

**2.** Describe and interpret any symmetry exhibited by the graph in Guided Practice 1.

**2 Interpret Extrema and End Behavior**  Interpreting a graph also involves estimating and interpreting where the function is increasing, decreasing, positive, or negative, and where the function has any extreme values, either high or low.

**KeyConcepts  Positive, Negative, Increasing, Decreasing, Extrema, and End Behavior**

A function is **positive** where its graph lies *above* the $x$-axis, and **negative** where its graph lies *below* the $x$-axis.

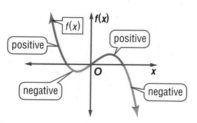

A function is **increasing** where the graph goes *up* and **decreasing** where the graph goes *down* when viewed from left to right.

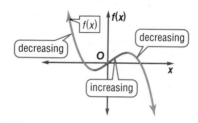

The points shown are the locations of relatively high or low function values called **extrema**. Point $A$ is a **relative minimum**, since no other nearby points have a lesser $y$-coordinate. Point $B$ is a **relative maximum**, since no other nearby points have a greater $y$-coordinate.

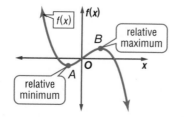

**End behavior** describes the values of a function at the positive and negative extremes in its domain.

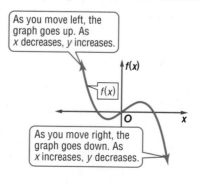

As you move left, the graph goes up. As $x$ decreases, $y$ increases.

As you move right, the graph goes down. As $x$ increases, $y$ decreases.

**Real-World**Link

The first successful commercially sold portable video game system was released in 1989 and sold for $120.

**Source:** *PCWorld*

**Study**Tip

Constant A function is *constant* where the graph does not go up or down as the graph is viewed from left to right.

**VIDEO GAMES** U.S. retail sales of video games from 2000 to 2009 can be modeled by the function graphed at the right. Estimate and interpret where the function is positive, negative, increasing, and decreasing, the *x*-coordinates of any relative extrema, and the end behavior of the graph.

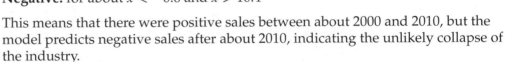

U.S. Video Games Sales

Years Since 2000

**Positive:** between about $x = -0.6$ and $x = 10.4$

**Negative:** for about $x < -0.6$ and $x > 10.4$

This means that there were positive sales between about 2000 and 2010, but the model predicts negative sales after about 2010, indicating the unlikely collapse of the industry.

**Increasing:** for about $x < 1.5$ and between about $x = 3$ and $x = 8$

**Decreasing:** between about $x = 2$ and $x = 3$ and for about $x > 8$

This means that sales increased from about 2000 to 2002, decreased from 2002 to 2003, increased from 2003 to 2008, and have been decreasing since 2008.

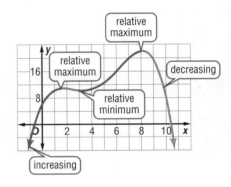

**Relative Maxima:** at about $x = 1.5$ and $x = 8$

**Relative Minima:** at about $x = 3$

The extrema of the graph indicate that the industry experienced two relative peaks in sales during this period: one around 2002 of approximately $10.5 billion and another around 2008 of approximately $22 billion. A relative low of $10 billion in sales came in about 2003.

**End Behavior:**
As *x* increases or decreases, the value of *y* decreases.

The end behavior of the graph indicates negative sales several years prior to 2000 and several years after 2009, which is unlikely. This graph appears to only model sales well between 2000 and 2009 and can only be used to predict sales in 2010.

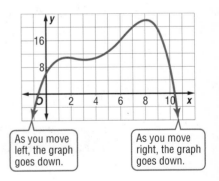

**Guided**Practice

**3.** Estimate and interpret where the function graphed in Guided Practice 1 is positive, negative, increasing, or decreasing, the *x*-coordinate of any relative extrema, and the end behavior of the graph.

## Check Your Understanding

**Examples 1–3** Identify the function graphed as *linear* or *nonlinear*. Then estimate and interpret the intercepts of the graph, any symmetry, where the function is positive, negative, increasing, and decreasing, the *x*-coordinate of any relative extrema, and the end behavior of the graph.

**1.**
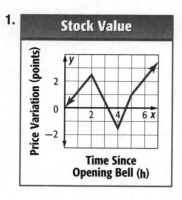

**2.**
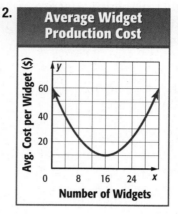

**3.**
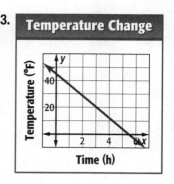

## Practice and Problem Solving

**Examples 1–3** Identify the function graphed as *linear* or *nonlinear*. Then estimate and interpret the intercepts of the graph, any symmetry, where the function is positive, negative, increasing, and decreasing, the *x*-coordinate of any relative extrema, and the end behavior of the graph.

**4.**
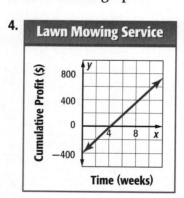

**5.**
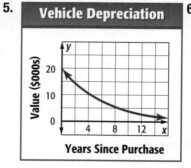

**6.**
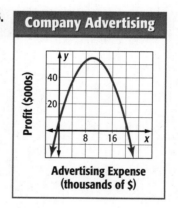

**7.**

**8.**
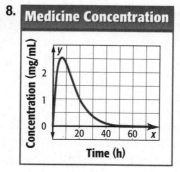

**9.**
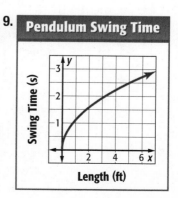

**10. FERRIS WHEEL** At the beginning of a Ferris wheel ride, a passenger cart is located at the same height as the center of the wheel. The position $y$ in feet of this cart relative to the center $t$ seconds after the ride starts is given by the function graphed at the right. Identify and interpret the key features of the graph. (*Hint:* Look for a pattern in the graph to help you describe its end behavior.)

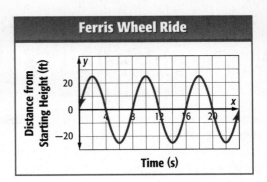

**Ferris Wheel Ride**

**Sketch a graph of a function that could represent each situation. Identify and interpret the intercepts of the graph, where the graph is increasing and decreasing, and any relative extrema.**

**11.** the height of a corn plant from the time the seed is planted until it reaches maturity 120 days later

**12.** the height of a football from the time it is punted until it reaches the ground 2.8 seconds later

**13.** the balance due on a car loan from the date the car was purchased until it was sold 4 years later

**Sketch graphs of functions with the following characteristics.**

**14.** The graph is linear with an $x$-intercept at $-2$. The graph is positive for $x < -2$, and negative for $x > -2$.

**15.** A nonlinear graph has $x$-intercepts at $-2$ and $2$ and a $y$-intercept at $-4$. The graph has a relative minimum of 4 at $x = 0$. The graph is decreasing for $x < 0$ and increasing for $x > 0$.

**16.** A nonlinear graph has a $y$-intercept at 2, but no $x$-intercepts. The graph is positive and increasing for all values of $x$.

**17.** A nonlinear graph has $x$-intercepts at $-8$ and $-2$ and a $y$-intercept at 3. The graph has relative minimums at $x = -3$ and $x = 6$ and a relative maximum at $x = 2$. The graph is positive for $x < -8$ and $x > -1$ and negative between $x = -8$ and $x = -1$. As $x$ decreases, $y$ increases and as $x$ increases, $y$ increases.

**H.O.T. Problems** Use Higher-Order Thinking Skills

**18. ERROR ANALYSIS** Katara thinks that all linear functions have exactly one $x$-intercept. Desmond thinks that a linear function can have at most one $x$-intercept. Is either of them correct? Explain your reasoning.

**19. CHALLENGE** Describe the end behavior of the graph shown.

**20. REASONING** Determine whether the following statement is *true* or *false*. Explain.

*Functions have at most one y-intercept.*

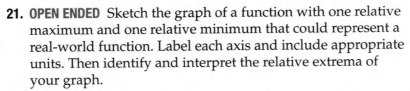

**21. OPEN ENDED** Sketch the graph of a function with one relative maximum and one relative minimum that could represent a real-world function. Label each axis and include appropriate units. Then identify and interpret the relative extrema of your graph.

**22. WRITING IN MATH** Describe how you would identify the key features of a graph described in this lesson using a table of values for a function.

# Spreadsheet Lab
# Descriptive Modeling

When using numbers to model a real-world situation, it is often helpful to have a metric. A **metric** is a rule for assigning a number to some characteristic or attribute. For example, teachers use metrics to determine grades. Each teacher determines an appropriate metric for assessing a student's performance and assigning a grade.

You can use a spreadsheet to calculate different metrics.

## Activity

**Dorrie wants to buy a house. She has the following expenses: rent of $650, credit card monthly bills of $320, a car payment of $410, and a student loan payment of $115. Dorrie has a yearly salary of $46,500. Use a spreadsheet to find Dorrie's debt-to-income ratio.**

**Step 1** Enter Dorrie's debts in column B.

**Step 2** Add her debts using a function in cell B6. Go to Insert and then Function. Then choose Sum. The sum of 1495 appears in B6.

**Step 3** Now insert Dorrie's salary in column C. Remember to find her monthly salary by dividing the yearly salary by 12.

A mortgage company will use the debt-to-income ratio as a metric to determine if Dorrie qualifies for a loan. The **debt-to-income ratio** is calculated as *how much she owes per month* divided by *how much she earns each month*.

**Step 4** Enter a formula to find the debt-to-income ratio in cell C6. In the formula bar, enter =B6/C2.

The ratio of about 0.39 appears. An ideal ratio would be 0.36 or less. A ratio higher than 0.36 would cause an increased interest rate or may require a higher down payment.

The spreadsheet shows a debt-to-income ratio of about 0.39. Dorrie should try to eliminate or reduce some debts or try to earn more money in order to lower her debt-to-income ratio.

### Lab 2-6 B Spreadsheet.xls

| | A | B | C |
|---|---|---|---|
| 1 | Type of Debt | Expenses | Salary |
| 2 | Rent | 650 | 3875 |
| 3 | Credit Cards | 320 | |
| 4 | Car Payment | 410 | |
| 5 | Student Loan | 115 | |
| 6 | | 1495 | 0.385806 |
| 7 | | | |

Sheet 1 / Sheet 2 / Sheet 3 /

## Exercises

1. How could Dorrie improve her debt-to-income ratio?

2. Another metric mortgage companies use is the ratio of monthly mortgage to total monthly income. An ideal ratio is 0.28. Using this metric, how much could Dorrie afford to pay for a mortgage each month?

3. How effective are each of these metrics as measures of whether Dorrie can afford to buy a house? Explain your reasoning.

4. **RESEARCH** Metrics are used to compare athletes. For example, ERAs are used to compare pitchers. Find a metric and evaluate its effectiveness for modeling. Compare it to other metrics, and then define your own metric.

# Algebra Lab
# Analyzing Linear Graphs

Analyzing a graph can help you learn about the relationship between two quantities. A **linear function** is a function for which the graph is a line. There are four types of linear graphs. Let's analyze each type.

## Activity 1    Line that Slants Up

**Analyze the function graphed at the right.**

**a. Describe the domain, range, and end behavior.**

**b. Describe the intercepts and any maximum or minimum points.**

**c. Identify where the function is positive, negative, increasing, and decreasing.**

**d. Describe any symmetry.**

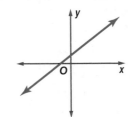

**a.** The domain and range are all real numbers. As you move left, the graph goes down. So as $x$ decreases, $y$ decreases. As you move right, the graph goes up. So as $x$ increases, $y$ increases.

**b.** There is one $x$-intercept and one $y$-intercept. There are no maximum or minimum points.

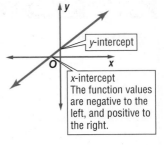

**c.** The function value is 0 at the $x$-intercept. The function values are negative to the left of the $x$-intercept and positive to the right. The function goes up from left to right, so it is increasing on the entire domain.

**d.** The graph has no symmetry.

Lines that slant down from left to right have some different key features.

## Activity 2    Line that Slants Down

**Analyze the function graphed at the right.**

**a. Describe the domain, range, and end behavior.**

**b. Describe the intercepts and any maximum or minimum points.**

**c. Identify where the function is positive, negative, increasing, and decreasing.**

**d. Describe any symmetry.**

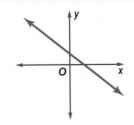

**a.** The domain and range are all real numbers. As you move left, the graph goes up. So as $x$ decreases, $y$ increases. As you move right, the graph goes down. So as $x$ increases, $y$ decreases.

**b.** There is one $x$-intercept and one $y$-intercept. There are no maximum or minimum points.

**c.** The function values are positive to the left of the $x$-intercept and negative to the right.
The function goes down from left to right, so it is decreasing on the entire domain.

**d.** The graph has no symmetry.

*(continued on the next page)*

Horizontal lines represent special functions called **constant functions**.

## Activity 3   Horizontal Line

**Analyze the function graphed at the right.**

a. The domain is all real numbers, and the range is one value. As you move left or right, the graph stays constant. So as $x$ decreases or increases, $y$ is constant.

b. The graph does not intersect the $x$-axis, so there is no $x$-intercept. The graph has one $y$-intercept. There are no maximum or minimum points.

c. The function values are all positive. The function is constant on the entire domain.

d. The graph is symmetric about any vertical line.

Vertical lines represent linear relations that are *not* functions.

## Activity 4   Vertical Line

**Analyze the relation graphed at the right.**

a. The domain is one value, and the range is all real numbers. This relation is not a function. Because you cannot move left or right on the graph, there is no end behavior.

b. There is one $x$-intercept and no $y$-intercept. There are no maximum or minimum points.

c. The $y$-values are positive above the $x$-axis and negative below. Because you cannot move left or right on the graph, the relation is neither increasing nor decreasing.

d. The graph is symmetric about itself.

## Analyze the Results

1. Compare and contrast the key features of lines that slant up and lines that slant down.

2. How would the key features of a horizontal line below the $x$-axis differ from the features of a line above the $x$-axis?

3. Consider lines that pass through the origin.

   a. How do the key features of a line that slants up and passes through the origin compare to the key features of the line in Activity 1?

   b. Compare the key features of a line that slants down and passes through the origin to the key features of the line in Activity 2.

   c. Describe a horizontal line that passes through the origin and a vertical line that passes through the origin. Compare their key features to those of the lines in Activities 3 and 4.

4. Place a pencil on a coordinate plane to represent a line. Move the pencil to represent different lines and evaluate each conjecture.

   a. *True* or *false*: A line can have more than one $x$-intercept.

   b. *True* or *false*: If the end behavior of a line is that as $x$ increases, $y$ increases, then the function values are increasing over the entire domain.

   c. *True* or *false*: Two different lines can have the same $x$- and $y$-intercepts.

## Sketch a linear graph that fits each description.

5. as $x$ increases, $y$ decreases

6. one $x$-intercept and one $y$-intercept

7. has symmetry

8. is not a function

# Regression and Median-Fit Lines

| | Then | | Now | | Why? |
|---|---|---|---|---|---|

**:· Then**

● You used lines of fit and scatter plots to evaluate trends and make predictions.

**:· Now**

**1** Write equations of best-fit lines using linear regression.

**2** Write equations of median-fit lines.

**:· Why?**

● The table shows the total attendance, in millions of people, at the Minnesota State Fair from 2005 to 2009. You can use a graphing calculator to find the equation of a *best-fit line* and use it to make predictions about future attendance at the fair.

| Year | Attendance (millions) |
|------|----------------------|
| 2005 | 1.633 |
| 2006 | 1.681 |
| 2007 | 1.682 |
| 2008 | 1.693 |
| 2009 | 1.790 |

**NewVocabulary**
best-fit line
linear regression
correlation coefficient
residual
median-fit line

**1 Best-Fit Lines** You have learned how to find and write equations for lines of fit by hand. Many calculators use complex algorithms that find a more precise line of fit called the **best-fit line**. One algorithm is called **linear regression**.

Your calculator may also compute a number called the **correlation coefficient**. This number will tell you if your correlation is positive or negative and how closely the equation is modeling the data. The closer the correlation coefficient is to 1 or −1, the more closely the equation models the data.

● **Real-World Example 1** Best-Fit Line

**MOVIES** The table shows the amount of money made by movies in the United States. Use a graphing calculator to write an equation for the best-fit line for that data.

| Year | 2000 | 2001 | 2002 | 2003 | 2004 | 2005 | 2006 | 2007 | 2008 | 2009 |
|------|------|------|------|------|------|------|------|------|------|------|
| Income ($ billion) | 7.48 | 8.13 | 9.19 | 9.35 | 9.27 | 8.95 | 9.25 | 9.65 | 9.85 | 10.21 |

Before you begin, make sure that your Diagnostic setting is on. You can find this under the **CATALOG** menu. Press **D** and then scroll down and click **DiagnosticOn**. Then press ENTER.

**Step 1** Enter the data by pressing STAT and selecting the **Edit** option. Let the year 2000 be represented by 0. Enter the years since 2000 into List 1 (**L1**). These will represent the *x*-values. Enter the income ($ billion) into List 2 (**L2**). These will represent the *y*-values.

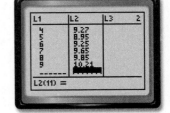

**Step 2** Perform the regression by pressing STAT and selecting the **CALC** option. Scroll down to **LinReg (ax+b)** and press ENTER twice.

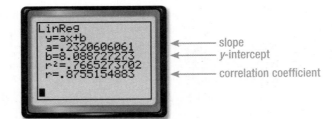

- slope
- *y*-intercept
- correlation coefficient

**Step 3** Write the equation of the regression line by rounding the $a$ and $b$ values on the screen. The form that we chose for the regression was $ax + b$, so the equation is $y = 0.23x + 8.09$. The correlation coefficient is about 0.8755, which means that the equation models the data fairly well.

▶ **Guided**Practice

**Write an equation of the best-fit line for the data in each table. Name the correlation coefficient. Round to the nearest ten-thousandth. Let $x$ be the number of years since 2003.**

**1A. HOCKEY** The table shows the number of goals of leading scorers for the Mustang Girls Hockey Team.

| Year | 2003 | 2004 | 2005 | 2006 | 2007 | 2008 | 2009 | 2010 |
|------|------|------|------|------|------|------|------|------|
| Goals | 30 | 23 | 41 | 35 | 31 | 43 | 33 | 45 |

**1B. HOCKEY** The table gives the number of goals scored by the team each season.

| Year | 2003 | 2004 | 2005 | 2006 | 2007 | 2008 | 2009 | 2010 |
|------|------|------|------|------|------|------|------|------|
| Goals | 63 | 44 | 55 | 63 | 81 | 85 | 93 | 84 |

We know that not all of the points will lie on the best-fit line. The difference between an observed $y$-value and its predicted $y$-value (found on the best-fit line) is called a **residual**. Residuals measure how much the data deviate from the regression line. When residuals are plotted on a scatter plot they can help to assess how well the best-fit line describes the data. If the best-fit line is a good fit, there is no pattern in the residual plot.

🌐 **Real-World Example 2** Graph and Analyze a Residual Plot

**HOCKEY** Graph and analyze the residual plot for the data for Guided Practice 1A. Determine if the best-fit line models the data well.

After calculating the least-squares regression line in Guided Practice 1A, you can obtain the residual plot of the data. Turn on **Plot2** under the STAT PLOT menu and choose ⠇⠂⠄. Use **L1** for the **Xlist** and **RESID** for the **Ylist**. You can obtain **RESID** by pressing ⎡2nd⎤ [STAT] and selecting **RESID** from the list of names. Graph the scatter plot of the residuals by pressing ⎡ZOOM⎤ and choosing **ZoomStat**.

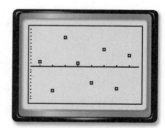

[0, 8] scl: 1 by [−10, 10] scl: 2

The residuals appear to be randomly scattered and centered about the line $y = 0$. Thus, the best-fit line seems to model the data well.

▶ **Guided**Practice

**2. UNEMPLOYMENT** Graph and analyze the residual plot for the following data comparing graduation rates and unemployment rates.

| Graduation Rate | 73 | 85 | 64 | 81 | 68 | 82 |
|-----------------|-----|-----|-----|-----|-----|-----|
| Unemployment Rate | 6.9 | 4.1 | 3.2 | 5.5 | 4.3 | 5.1 |

A residual is positive when the observed value is above the line, negative when the observed value is below the line, and zero when it is on the line. One common measure of goodness of fit is the sum of squared vertical distances from the points to the line. The best-fit line, which is also called the *least-squares regression line*, minimizes the sum of the squares of those distances.

We can use points on the best-fit line to estimate values that are not in the data. Recall that when we estimate values that are between known values, this is called *linear interpolation*. When we estimate a number outside of the range of the data, it is called *linear extrapolation*.

---

### Real-World Example 3  Use Interpolation and Extrapolation

**PAINTBALL**  **The table shows the points received by the top ten paintball teams at a tournament. Estimate how many points the 20th-ranked team received.**

| Rank | 1 | 2 | 3 | 4 | 5 | 6 | 7 | 8 | 9 | 10 |
|------|-----|-----|-----|-----|-----|-----|-----|-----|-----|-----|
| Score | 100 | 89 | 96 | 99 | 97 | 98 | 78 | 70 | 64 | 80 |

Write an equation of the best-fit line for the data. Then extrapolate to find the missing value.

**Step 1** Enter the data from the table into the lists. Let the ranks be the *x*-values and the scores be the *y*-values. Then graph the scatter plot.

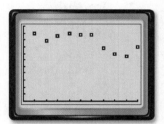

[0, 10] scl: 1 by [0, 110] scl: 10

**Step 2** Perform the linear regression using the data in the lists. Find the equation of the best-fit line.

The equation is about $y = -3.32x + 105.3$.

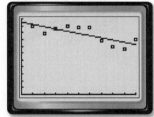

[0, 10] scl: 1 by [0, 110] scl: 10

**Step 3** Graph the best-fit line. Press Y= VARS and choose **Statistics**. From the EQ menu, choose **RegEQ**. Then press GRAPH.

**Step 4** Use the graph to predict the points that the 20th-ranked team received. Change the viewing window to include the *x*-value to be evaluated. Press 2nd [CALC] ENTER 20 ENTER to find that when $x = 20$, $y \approx 39$. It is estimated that the 20th ranked team received 39 points.

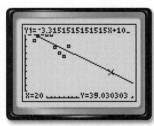

[0, 25] scl: 1 by [0, 110] scl: 1

**Study**Tip

Median-Fit Line
The median-fit line is computed using a different algorithm than linear regression.

► GuidedPractice

**ONLINE GAMES** Use linear interpolation to estimate the percent of Americans that play online games for the following ages.

| Age | 15 | 20 | 30 | 40 | 50 |
|---|---|---|---|---|---|
| Percent | 81 | 54 | 37 | 29 | 25 |

**Source:** Pew Internet & American Life Survey

**3A.** 35 years

**3B.** 18 years

## 2 Median-Fit Lines
A second type of fit line that can be found using a graphing calculator is a **median-fit line**. The equation of a median-fit line is calculated using the medians of the coordinates of the data points.

### Example 4  Median-Fit Line

**PAINTBALL** Find and graph the equation of a median-fit line for the data in Example 3. Then predict the score of the 15th ranked team.

**Step 1** Reenter the data if it is not in the lists. Clear the **Y=** list and graph the scatter plot.

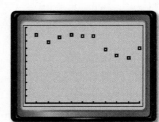

[0, 10] scl: 1 by [0, 110] scl: 10

**Step 2** To find the median-fit equation, press the [STAT] key and select the **CALC** option. Scroll down to the **Med-Med** option and press [ENTER]. The value of $a$ is the slope, and the value of $b$ is the $y$-intercept.

The equation for the median-fit line is about $y = -3.71x + 108.26$.

**Step 3** Copy the equation to the **Y=** list and graph. Use the **value** option to find the value of $y$ when $x = 15$.

The 15th place team scored about 53 points.

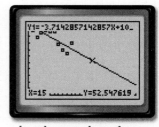

[0, 25] scl: 1 by [0, 110] scl: 1

Notice that the equations for the regression line and the median-fit line are very similar.

► GuidedPractice

**4.** Use the data from Guided Practice 3 and a median-fit line to estimate the numbers of 18- and 35-year-olds who play online games. Compare these values with the answers from the regression line.

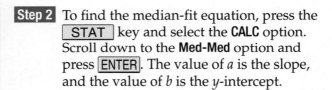

### Real-WorldLink

Paintball is more popular with 12- to 17-year-olds than any other age group. In a recent year, 3,649,000 teenagers participated in paintball while 2,195,000 18- to 24-year-olds participated.

**Source:** *Statistical Abstract* of the *United States*

Alan Thornton/Stone/Getty Images

**Examples 1, 2** **1. POTTERY** A local university is keeping track of the number of art students who use the pottery studio each day.

| Day | 1 | 2 | 3 | 4 | 5 | 6 | 7 |
|---|---|---|---|---|---|---|---|
| Students | 10 | 15 | 18 | 15 | 13 | 19 | 20 |

    **a.** Write an equation of the regression line and find the correlation coefficient.

    **b.** Graph the residual plot and determine if the regression line models the data well.

**Example 3** **2. COMPUTERS** The table below shows the percent of Americans with a broadband connection at home in a recent year. Use linear extrapolation and a regression equation to estimate the percentage of 60-year-olds with broadband at home.

| Age | 25 | 30 | 35 | 40 | 45 | 50 |
|---|---|---|---|---|---|---|
| Percent | 40 | 42 | 36 | 35 | 36 | 32 |

**Example 4** **3. VACATION** The Smiths want to rent a house on the lake that sleeps eight people. The cost of the house per night is based on how close it is to the water.

| Distance from Lake (mi) | 0.0 (houseboat) | 0.3 | 0.5 | 1.0 | 1.25 | 1.5 | 2.0 |
|---|---|---|---|---|---|---|---|
| Price/Night ($) | 785 | 325 | 250 | 200 | 150 | 140 | 100 |

    **a.** Find and graph an equation for the median-fit line.

    **b.** What would you estimate is the cost of a rental 1.75 miles from the lake?

## Practice and Problem Solving

**Example 1** **Write an equation of the regression line for the data in each table. Then find the correlation coefficient.**

    **4. SKYSCRAPERS** The table ranks the ten tallest buildings in the world.

| Rank | 1 | 2 | 3 | 4 | 5 | 6 | 7 | 8 | 9 | 10 |
|---|---|---|---|---|---|---|---|---|---|---|
| Stories | 101 | 88 | 110 | 88 | 88 | 80 | 69 | 102 | 78 | 70 |

    **5. MUSIC** The table gives the number of annual violin auditions held by a youth symphony each year since 2004. Let $x$ be the number of years since 2004.

| Year | 2004 | 2005 | 2006 | 2007 | 2008 | 2009 | 2010 |
|---|---|---|---|---|---|---|---|
| Auditions | 22 | 19 | 25 | 37 | 32 | 35 | 42 |

**Example 2** **6. RETAIL** The table gives the sales at a clothing chain since 2004. Let $x$ be the number of years since 2004.

| Year | 2004 | 2005 | 2006 | 2007 | 2008 | 2009 |
|---|---|---|---|---|---|---|
| Sales (Millions of Dollars) | 6.84 | 7.6 | 10.9 | 15.4 | 17.6 | 21.2 |

    **a.** Write an equation of the regression line.

    **b.** Graph and analyze the residual plot.

**Examples 3, 4** **7. MARATHON** The number of entrants in the Boston Marathon every five years since 1975 is shown. Let $x$ be the number of years since 1975.

| Year | 1975 | 1980 | 1985 | 1990 | 1995 | 2000 | 2005 | 2010 |
|---|---|---|---|---|---|---|---|---|
| Entrants | 2395 | 5417 | 5594 | 9412 | 9416 | 17,813 | 20,453 | 26,735 |

   **a.** Find an equation for the median-fit line.

   **b.** According to the equation, how many entrants were there in 2003?

**8. CAMPING** A campground keeps a record of the number of campsites rented the week of July 4 for several years. Let $x$ be the number of years since 2000.

| Year | 2002 | 2003 | 2004 | 2005 | 2006 | 2007 | 2008 | 2009 | 2010 |
|---|---|---|---|---|---|---|---|---|---|
| Sites Rented | 34 | 45 | 42 | 53 | 58 | 47 | 57 | 65 | 59 |

   **a.** Find an equation for the regression line.

   **b.** Predict the number of campsites that will be rented in 2012.

   **c.** Predict the number of campsites that will be rented in 2020.

**9. ICE CREAM** An ice cream company keeps a count of the tubs of chocolate ice cream delivered to each of their stores in a particular area.

   **a.** Find an equation for the median-fit line.

   **b.** Graph the points and the median-fit line.

   **c.** How many tubs would be delivered to a 1500-square-foot store? a 5000-square-foot store?

| Store Size (ft²) | 2100 | 2225 | 3135 | 3569 | 4587 |
|---|---|---|---|---|---|
| Tubs (hundreds) | 110 | 102 | 215 | 312 | 265 |

**10. FINANCIAL LITERACY** The prices of the eight top-selling brands of jeans at Jeanie's Jeans are given in the table below.

| Sales Rank | 1 | 2 | 3 | 4 | 5 | 6 | 7 | 8 |
|---|---|---|---|---|---|---|---|---|
| Price ($) | 43 | 44 | 50 | 61 | 64 | 135 | 108 | 78 |

   **a.** Find the equation for the regression line.

   **b.** According to the equation, what would be the price of a pair of the 12th best-selling brand?

   **c.** Is this a reasonable prediction? Explain.

**11. STATE FAIRS** Refer to the beginning of the lesson.

   **a.** Graph a scatter plot of the data, where $x = 1$ represents 2005. Then find and graph the equation for the best-fit line.

   **b.** Graph and analyze the residual plot.

   **c.** Predict the total attendance in 2020.

**12. FIREFIGHTERS** The table shows statistics from the U.S. Fire Administration.

  **a.** Find an equation for the median-fit line.

  **b.** Graph the points and the median-fit line.

  **c.** Does the median-fit line give you an accurate picture of the number of firefighters? Explain.

| Age | Number of Firefighters |
|-----|------------------------|
| 18  | 40,919                 |
| 25  | 245,516                |
| 35  | 330,516                |
| 45  | 296,665                |
| 55  | 167,087                |
| 65  | 54,559                 |

**13. ATHLETICS** The table shows the number of participants in high school athletics.

| Year Since 1970 | 1 | 10 | 20 | 30 | 35 |
|-----------------|---|----|----|----|----|
| Athletes | 3,960,932 | 5,356,913 | 5,298,671 | 6,705,223 | 7,159,904 |

  **a.** Find an equation for the regression line.

  **b.** According to the equation, how many participated in 1988?

**14. ART** A count was kept on the number of paintings sold at an auction by the year in which they were painted. Let $x$ be the number of years since 1950.

| Year Painted | 1950 | 1955 | 1960 | 1965 | 1970 | 1975 |
|--------------|------|------|------|------|------|------|
| Paintings Solds | 8 | 5 | 25 | 21 | 9 | 22 |

  **a.** Find the equation for the linear regression line.

  **b.** How many paintings were sold that were painted in 1961?

  **c.** Is the linear regression equation an accurate model of the data? Explain why or why not.

## H.O.T. Problems    Use Higher-Order Thinking Skills

**15. CHALLENGE** Below are the results of the World Superpipe Championships in 2008.

| Men | Score | Rank | Women | Score |
|-----|-------|------|-------|-------|
| Shaun White | 93.00 | 1 | Torah Bright | 96.67 |
| Mason Aguirre | 90.33 | 2 | Kelly Clark | 93.00 |
| Janne Korpi | 85.33 | 3 | Soko Yamaoka | 85.00 |
| Luke Mitrani | 85.00 | 4 | Ellery Hollingsworth | 79.33 |
| Keir Dillion | 81.33 | 5 | Sophie Rodriguez | 71.00 |

Find an equation of the regression line for each, and graph them on the same coordinate plane. Compare and contrast the men's and women's graphs.

**16. REASONING** For a class project, the scores that 10 randomly selected students earned on the first 8 tests of the school year are given. Explain how to find a line of best fit. Could it be used to predict the scores of other students? Explain your reasoning.

**17. OPEN ENDED** For 10 different people, measure their heights and the lengths of their heads from chin to top. Use these data to generate a linear regression equation and a median-fit equation. Make a prediction using both of the equations.

**18. ✎ WRITING IN MATH** How are lines of fit and linear regression similar? different?

# 6 Inverse Linear Functions

| ·· Then | ·· Now | ·· Why? |
|---|---|---|
| • You represented relations as tables, graphs, and mappings. | • **1** Find the inverse of a relation. <br> • **2** Find the inverse of a linear function. | • Randall is writing a report on Santiago, Chile, and he wants to include a brief climate analysis. He found a table of temperatures recorded in degrees Celsius. He knows that a formula for converting degrees Fahrenheit to degrees Celsius is $C(x) = \frac{5}{9}(x - 32)$. He will need to find the *inverse* function to convert from degrees Celsius to degrees Fahrenheit. |

| Average Temp (°C) | | |
|---|---|---|
| **Month** | **Min** | **Max** |
| Jan | 12 | 29 |
| March | 9 | 27 |
| May | 5 | 18 |
| July | 3 | 15 |
| Sept | 6 | 29 |
| Nov | 9 | 26 |

### NewVocabulary
inverse relation
inverse function

**1 Inverse Relations** An **inverse relation** is the set of ordered pairs obtained by exchanging the $x$-coordinates with the $y$-coordinates of each ordered pair in a relation. If (5, 3) is an ordered pair of a relation, then (3, 5) is an ordered pair of the inverse relation.

> **KeyConcept** Inverse Relations
>
> **Words** If one relation contains the element $(a, b)$, then the inverse relation will contain the element $(b, a)$.
>
> **Example** $A$ and $B$ are inverse relations.
>
> | $A$ | | $B$ |
> |---|---|---|
> | $(-3, -16)$ | ⟶ | $(-16, -3)$ |
> | $(-1, 4)$ | ⟶ | $(4, -1)$ |
> | $(2, 14)$ | ⟶ | $(14, 2)$ |
> | $(5, 32)$ | ⟶ | $(32, 5)$ |

Notice that the domain of a relation becomes the range of its inverse, and the range of the relation becomes the domain of its inverse.

**PT**

### Example 1 Inverse Relations

**Find the inverse of each relation.**

**a.** {(4, −10), (7, −19), (−5, 17), (−3, 11)}

To find the inverse, exchange the coordinates of the ordered pairs.

$(4, -10) \rightarrow (-10, 4)$        $(-5, 17) \rightarrow (17, -5)$
$(7, -19) \rightarrow (-19, 7)$        $(-3, 11) \rightarrow (11, -3)$

The inverse is {(−10, 4), (−19, 7), (17, −5), (11, −3)}.

**b.**

| x | −4 | −1 | 5 | 9 |
|---|---|---|---|---|
| y | −13 | −8.5 | 0.5 | 6.5 |

Write the coordinates as ordered pairs. Then exchange the coordinates of each pair.

$(-4, -13) \rightarrow (-13, -4)$        $(5, 0.5) \rightarrow (0.5, 5)$
$(-1, -8.5) \rightarrow (-8.5, -1)$        $(9, 6.5) \rightarrow (6.5, 9)$

The inverse is {(−13, −4), (−8.5, −1), (0.5, 5), (6.5, 9)}.

**1A.** {(−6, 8), (−15, 11), (9, 3), (0, 6)}

**1B.**

| x | −10 | −4 | −3 | 0 |
|---|-----|----|----|----|
| y | 5 | 11 | 12 | 15 |

The graphs of relations can be used to find and graph inverse relations.

**Example 2  Graph Inverse Relations**

**Graph the inverse of the relation.**

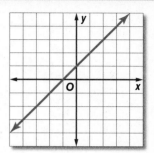

**StudyTip**

Graphing Inverses  Only two points are necessary to graph the inverse of a line, but several should be used to avoid possible error.

The graph of the relation passes through the points at (−4, −3), (−2, −1), (0, 1), (2, 3), and (3, 4). To find points through which the graph of the inverse passes, exchange the coordinates of the ordered pairs. The graph of the inverse passes through the points at (−3, −4), (−1, −2), (1, 0), (3, 2), and (4, 3). Graph these points and then draw the line that passes through them.

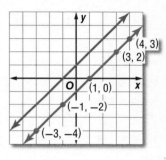

**GuidedPractice**

**Graph the inverse of each relation.**

**2A.**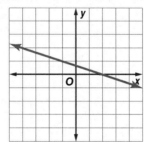

**2B.**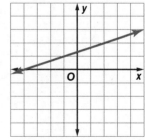

The graphs from Example 2 are graphed on the right with the line $y = x$. Notice that the graph of an inverse is the graph of the original relation reflected in the line $y = x$. For every point $(x, y)$ on the graph of the original relation, the graph of the inverse will include the point $(y, x)$.

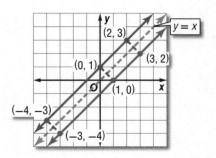

**2 Inverse Functions**  A linear relation that is described by a function has an **inverse function** that can generate ordered pairs of the inverse relation The inverse of the linear function $f(x)$ can be written as $f^{-1}(x)$ and is read $f$ of $x$ inverse or *the inverse of f of x.*

> **KeyConcept** Finding Inverse Functions

To find the inverse function $f^{-1}(x)$ of the linear function $f(x)$, complete the following steps.

**Step 1** Replace $f(x)$ with $y$ in the equation for $f(x)$.

**Step 2** Interchange $y$ and $x$ in the equation.

**Step 3** Solve the equation for $y$.

**Step 4** Replace $y$ with $f^{-1}(x)$ in the new equation.

> **Example 3** Find Inverse Linear Functions

**Find the inverse of each function.**

**a.** $f(x) = 4x - 8$

| Step 1 | $f(x) = 4x - 8$ | Original equation |
| | $y = 4x - 8$ | Replace $f(x)$ with $y$. |
| Step 2 | $x = 4y - 8$ | Interchange $y$ and $x$. |
| Step 3 | $x + 8 = 4y$ | Add 8 to each side. |
| | $\dfrac{x+8}{4} = y$ | Divide each side by 4. |
| Step 4 | $\dfrac{x+8}{4} = f^{-1}(x)$ | Replace $y$ with $f^{-1}(x)$. |

The inverse of $f(x) = 4x - 8$ is $f^{-1}(x) = \dfrac{x+8}{4}$ or $f^{-1}(x) = \frac{1}{4}x + 2$.

**CHECK** Graph both functions and the line $y = x$ on the same coordinate plane. $f^{-1}(x)$ appears to be the reflection of $f(x)$ in the line $y = x$. ✓

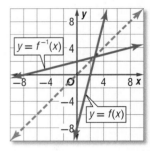

**WatchOut!**

Notation The $-1$ in $f^{-1}(x)$ is *not* an exponent.

**b.** $f(x) = -\frac{1}{2}x + 11$

| Step 1 | $f(x) = -\frac{1}{2}x + 11$ | Original equation |
| | $y = -\frac{1}{2}x + 11$ | Replace $f(x)$ with $y$. |
| Step 2 | $x = -\frac{1}{2}y + 11$ | Interchange $y$ and $x$. |
| Step 3 | $x - 11 = -\frac{1}{2}y$ | Subtract 11 from each side. |
| | $-2(x - 11) = y$ | Multiply each side by $-2$. |
| | $-2x + 22 = y$ | Distributive Property |
| Step 4 | $-2x + 22 = f^{-1}(x)$ | Replace $y$ with $f^{-1}(x)$. |

The inverse of $f(x) = -\frac{1}{2}x + 11$ is $f^{-1}(x) = -2x + 22$.

▶ **Guided Practice**

**3A.** $f(x) = 4x - 12$      **3B.** $f(x) = \frac{1}{3}x + 7$

**TEMPERATURE** Refer to the beginning of the lesson. Randall wants to convert the temperatures from degrees Celsius to degrees Fahrenheit.

**a. Find the inverse function $C^{-1}(x)$.**

| Step 1 | $C(x) = \frac{5}{9}(x - 32)$ | Original equation |
| | $y = \frac{5}{9}(x - 32)$ | Replace $C(x)$ with $y$. |
| Step 2 | $x = \frac{5}{9}(y - 32)$ | Interchange $y$ and $x$. |
| Step 3 | $\frac{9}{5}x = y - 32$ | Multiply each side by $\frac{9}{5}$. |
| | $\frac{9}{5}x + 32 = y$ | Add 32 to each side. |
| Step 4 | $\frac{9}{5}x + 32 = C^{-1}(x)$ | Replace $y$ with $C^{-1}(x)$. |

The inverse function of $C(x)$ is $C^{-1}(x) = \frac{9}{5}x + 32$.

**b. What do $x$ and $C^{-1}(x)$ represent in the context of the inverse function?**

$x$ represents the temperature in degrees Celsius. $C^{-1}(x)$ represents the temperature in degrees Fahrenheit.

**c. Find the average temperatures for July in degrees Fahrenheit.**

The average minimum and maximum temperatures for July are 3° C and 15° C, respectively. To find the average minimum temperature, find $C^{-1}(3)$.

$C^{-1}(x) = \frac{9}{5}x + 32$    Original equation

$C^{-1}(3) = \frac{9}{5}(3) + 32$    Substitute 3 for $x$.

$\quad\quad = 37.4$    Simplify.

To find the average maximum temperature, find $C^{-1}(15)$.

$C^{-1}(x) = \frac{9}{5}x + 32$    Original equation

$C^{-1}(15) = \frac{9}{5}(15) + 32$    Substitute 15 for $x$.

$\quad\quad = 59$    Simplify.

The average minimum and maximum temperatures for July are 37.4° F and 59° F, respectively.

**Guided Practice**

**4. RENTAL CAR** Peggy rents a car for the day. The total cost $C(x)$ in dollars is given by $C(x) = 19.99 + 0.3x$, where $x$ is the number of miles she drives.

**A.** Find the inverse function $C^{-1}(x)$.

**B.** What do $x$ and $C^{-1}(x)$ represent in the context of the inverse function?

**C.** How many miles did Peggy drive if her total cost was $34.99?

**Real-World** Link

The winter months in Chile occur during the summer months in the U.S. due to Chile's location in the southern hemisphere. The average daily high temperature of Santiago during its winter months is about 60° F.

**Source:** World Weather Information Service

**Example 1**    Find the inverse of each relation.

1. $\{(4, -15), (-8, -18), (-2, -16.5), (3, -15.25)\}$

2.

| $x$ | −3 | 0 | 1 | 6 |
|---|---|---|---|---|
| $y$ | 11.8 | 3.7 | 1 | −12.5 |

**Example 2**    Graph the inverse of each relation.

3.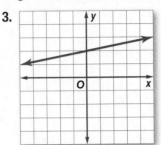

4.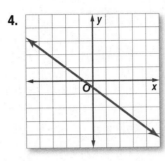

**Example 3**    Find the inverse of each function.

5. $f(x) = -2x + 7$

6. $f(x) = \frac{2}{3}x + 6$

**Example 4**    7. **TICKETS** Dwayne and his brother purchase season tickets to the Cleveland Crusaders games. The ticket package requires a one-time purchase of a personal seat license costing $1200 for two seats. A ticket to each game costs $70. The cost $C(x)$ in dollars for Dwayne for the first season is $C(x) = 600 + 70x$, where $x$ is the number of games Dwayne attends.

   a. Find the inverse function.

   b. What do $x$ and $C^{-1}(x)$ represent in the context of the inverse function?

   c. How many games did Dwayne attend if his total cost for the season was $950?

## Practice and Problem Solving

**Example 1**    Find the inverse of each relation.

8. $\{(-5, 13), (6, 10.8), (3, 11.4), (-10, 14)\}$

9. $\{(-4, -49), (8, 35), (-1, -28), (4, 7)\}$

10.

| $x$ | $y$ |
|---|---|
| −8 | −36.4 |
| −2 | −15.4 |
| 1 | −4.9 |
| 5 | 9.1 |
| 11 | 30.1 |

11.

| $x$ | $y$ |
|---|---|
| −3 | 7.4 |
| −1 | 4 |
| 1 | 0.6 |
| 3 | −2.8 |
| 5 | −6.2 |

**Example 2**    Graph the inverse of each relation.

12.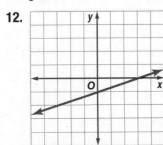

13.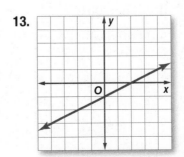

**Example 3**  **Find the inverse of each function.**

**14.** $f(x) = 25 + 4x$

**15.** $f(x) = 17 - \frac{1}{3}x$

**16.** $f(x) = 4(x + 17)$

**17.** $f(x) = 12 - 6x$

**18.** $f(x) = \frac{2}{5}x + 10$

**19.** $f(x) = -16 - \frac{4}{3}x$

**Example 4**

**20. DOWNLOADS** An online music subscription service allows members to download songs for $0.99 each after paying a monthly service charge of $3.99. The total monthly cost $C(x)$ of the service in dollars is $C(x) = 3.99 + 0.99x$, where $x$ is the number of songs downloaded.

  **a.** Find the inverse function.

  **b.** What do $x$ and $C^{-1}(x)$ represent in the context of the inverse function?

  **c.** How many songs were downloaded if a member's monthly bill is $27.75?

**21. LANDSCAPING** At the start of the mowing season, Chuck collects a one-time maintenance fee of $10 from his customers. He charges the Fosters $35 for each cut. The total amount collected from the Fosters in dollars for the season is $C(x) = 10 + 35x$, where $x$ is the number of times Chuck mows the Fosters' lawn.

  **a.** Find the inverse function.

  **b.** What do $x$ and $C^{-1}(x)$ represent in the context of the inverse function?

  **c.** How many times did Chuck mow the Fosters' lawn if he collected a total of $780 from them?

**Write the inverse of each equation in $f^{-1}(x)$ notation.**

**22.** $3y - 12x = -72$

**23.** $x + 5y = 15$

**24.** $-42 + 6y = x$

**25.** $3y + 24 = 2x$

**26.** $-7y + 2x = -28$

**27.** $3y - x = 3$

**Match each function with the graph of its inverse.**

**A.**
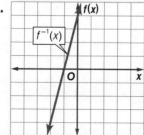

**B.**
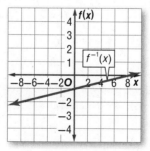

**C.**
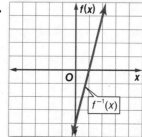

**D.**
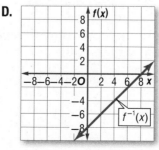

**28.** $f(x) = x + 4$

**29.** $f(x) = 4x + 4$

**30.** $f(x) = \frac{1}{4}x + 1$

**31.** $f(x) = \frac{1}{4}x - 1$

**Write an equation for the inverse function $f^{-1}(x)$ that satisfies the given conditions.**

**32.** slope of $f(x)$ is 7; graph of $f^{-1}(x)$ contains the point (13, 1)

**33.** graph of $f(x)$ contains the points (−3, 6) and (6, 12)

**34.** graph of $f(x)$ contains the point (10, 16); graph of $f^{-1}(x)$ contains the point (3, −16)

**35.** slope of $f(x)$ is 4; $f^{-1}(5) = 2$

**36. CELL PHONES** Mary Ann pays a monthly fee for her cell phone package which includes 700 minutes. She gets billed an additional charge for every minute she uses the phone past the 700 minutes. During her first month, Mary Ann used 26 additional minutes and her bill was $37.79. During her second month, Mary Ann used 38 additional minutes and her bill was $41.39.

   **a.** Write a function that represents the total monthly cost $C(x)$ of Mary Ann's cell phone package, where $x$ is the number of additional minutes used.

   **b.** Find the inverse function.

   **c.** What do $x$ and $C^{-1}(x)$ represent in the context of the inverse function?

   **d.** How many additional minutes did Mary Ann use if her bill for her third month was $48.89?

**37.**  🔹 **MULTIPLE REPRESENTATIONS** In this problem, you will explore the domain and range of inverse functions.

   **a. Algebraic** Write a function for the area $A(x)$ of the rectangle shown.

   **b. Graphical** Graph $A(x)$. Describe the domain and range of $A(x)$ in the context of the situation.

   **c. Algebraic** Write the inverse of $A(x)$. What do $x$ and $A^{-1}(x)$ represent in the context of the inverse function?

   **d. Graphical** Graph $A^{-1}(x)$. Describe the domain and range of $A^{-1}(x)$ in the context of the situation.

   **e. Logical** Determine the relationship between the domains and ranges of $A(x)$ and $A^{-1}(x)$.

8      Area = $A(x)$

$(x - 3)$

---

## H.O.T. Problems    Use Higher-Order Thinking Skills

**38. CHALLENGE** If $f(x) = 5x + a$ and $f^{-1}(10) = -1$, find $a$.

**39. CHALLENGE** If $f(x) = \frac{1}{a}x + 7$ and $f^{-1}(x) = 2x - b$, find $a$ and $b$.

**REASONING** Determine whether the following statements are *sometimes*, *always*, or *never* true. **Explain your reasoning.**

**40.** If $f(x)$ and $g(x)$ are inverse functions, then $f(a) = b$ and $g(b) = a$.

**41.** If $f(a) = b$ and $g(b) = a$, then $f(x)$ and $g(x)$ are inverse functions.

**42. OPEN ENDED** Give an example of a function and its inverse. Verify that the two functions are inverses by graphing the functions and the line $y = x$ on the same coordinate plane.

**43. WRITING IN MATH** Explain why it may be helpful to find the inverse of a function.

You can use patty paper to draw the graph of an inverse relation by reflecting the original graph in the line $y = x$.

### Activity Draw an Inverse

**Consider the graphs shown.**

**Step 1** Trace the graphs onto a square of patty paper, waxed paper, or tracing paper.

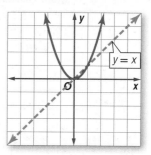

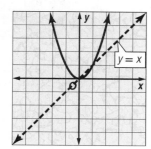

**Step 2** Flip the patty paper over and lay it on the original graph so that the traced $y = x$ is on the original $y = x$.

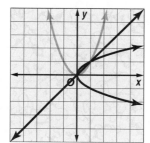

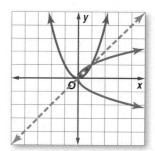

Notice that the result is the reflection of the graph in the line $y = x$ or the inverse of the graph.

## Analyze The Results

**1.** Is the graph of the original relation a function? Explain.

**2.** Is the graph of the inverse relation a function? Explain.

**3.** What are the domain and range of the original relation? of the inverse relation?

**4.** If the domain of the original relation is restricted to $D = \{x \mid x \geq 0\}$, is the inverse relation a function? Explain.

**5.** If the graph of a relation is a function, what can you conclude about the graph of its inverse?

**6.** **CHALLENGE** The vertical line test can be used to determine whether a relation is a function. Write a rule that can be used to determine whether a function has an inverse that is also a function.

# Rational Exponents

- You used the laws of exponents to find products and quotients of monomials.

- **1** Evaluate and rewrite expressions involving rational exponents.

- **2** Solve equations involving expressions with rational exponents.

- It's important to protect your skin with sunscreen to prevent damage. The sun protection factor (SPF) of a sunscreen indicates how well it protects you. Sunscreen with an SPF of $f$ absorbs about $p$ percent of the UV-B rays, where $p = 50f^{0.2}$.

 **NewVocabulary**
rational exponent
cube root
$n$th root
exponential equation

**1** **Rational Exponents** You know that an exponent represents the number of times that the base is used as a factor. But how do you evaluate an expression with an exponent that is not an integer like the one above? Let's investigate **rational exponents** by assuming that they behave like integer exponents.

$$\left(b^{\frac{1}{2}}\right)^2 = b^{\frac{1}{2}} \cdot b^{\frac{1}{2}} \qquad \text{Write as a multiplication expression.}$$
$$= b^{\frac{1}{2} + \frac{1}{2}} \qquad \text{Product of Powers}$$
$$= b^1 \text{ or } b \qquad \text{Simplify.}$$

Thus, $b^{\frac{1}{2}}$ is a number with a square equal to $b$. So $b^{\frac{1}{2}} = \sqrt{b}$.

---

**KeyConcept** $b^{\frac{1}{2}}$

| | |
|---|---|
| **Words** | For any nonnegative real number $b$, $b^{\frac{1}{2}} = \sqrt{b}$. |
| **Examples** | $16^{\frac{1}{2}} = \sqrt{16}$ or $4$ $\qquad\qquad$ $38^{\frac{1}{2}} = \sqrt{38}$ |

---

**Example 1** Radical and Exponential Forms

Write each expression in radical form, or write each radical in exponential form.

**a.** $25^{\frac{1}{2}}$
$25^{\frac{1}{2}} = \sqrt{25}$ $\quad$ Definition of $b^{\frac{1}{2}}$
$\qquad = 5$ $\qquad$ Simplify.

**b.** $\sqrt{18}$
$\sqrt{18} = 18^{\frac{1}{2}}$ $\quad$ Definition of $b^{\frac{1}{2}}$

**c.** $5x^{\frac{1}{2}}$
$5x^{\frac{1}{2}} = 5\sqrt{x}$ $\quad$ Definition of $b^{\frac{1}{2}}$

**d.** $\sqrt{8p}$
$\sqrt{8p} = (8p)^{\frac{1}{2}}$ $\quad$ Definition of $b^{\frac{1}{2}}$

▶ **GuidedPractice**

**1A.** $a^{\frac{1}{2}}$ $\qquad\qquad$ **1B.** $\sqrt{22}$ $\qquad\qquad$ **1C.** $(7w)^{\frac{1}{2}}$ $\qquad\qquad$ **1D.** $2\sqrt{x}$

You know that to find the square root of a number $a$ you find a number with a square of $a$. In the same way, you can find other roots of numbers. If $a^3 = b$, then $a$ is the **cube root** of $b$, and if $a^n = b$ for a positive integer $n$, then $a$ is an **$n$th root** of $b$.

**StudyTip**

Graphing Calculator You can use a graphing calculator to find $n$th roots. Enter $n$, then press MATH and choose $\sqrt[x]{\phantom{x}}$.

**KeyConcept** $n$th Root

| | |
|---|---|
| **Words** | For any real numbers $a$ and $b$ and any positive integer $n$, if $a^n = b$, then $a$ is an $n$th root of $b$. |
| **Symbols** | If $a^n = b$, then $\sqrt[n]{b} = a$. |
| **Example** | Because $2^4 = 16$, 2 is a fourth root of 16; $\sqrt[4]{16} = 2$. |

Since $3^2 = 9$ and $(-3)^2 = 9$, both 3 and $-3$ are square roots of 9. Similarly, since $2^4 = 16$ and $(-2)^4 = 16$, both 2 and $-2$ are fourth roots of 16. The positive roots are called *principal roots*. Radical symbols indicate principal roots, so $\sqrt[4]{16} = 2$.

**Example 2** $n$th roots

**Simplify.**

**a.** $\sqrt[3]{27}$

$$\sqrt[3]{27} = \sqrt[3]{3 \cdot 3 \cdot 3}$$
$$= 3$$

**b.** $\sqrt[5]{32}$

$$\sqrt[5]{32} = \sqrt[5]{2 \cdot 2 \cdot 2 \cdot 2 \cdot 2}$$
$$= 2$$

▶ **GuidedPractice**

**2A.** $\sqrt[3]{64}$

**2B.** $\sqrt[4]{10,000}$

Like square roots, $n$th roots can be represented by rational exponents.

$$\left(b^{\frac{1}{n}}\right)^n = \underbrace{b^{\frac{1}{n}} \cdot b^{\frac{1}{n}} \cdot \ldots \cdot b^{\frac{1}{n}}}_{n \text{ factors}}$$   Write as a multiplication expression.

$$= b^{\frac{1}{n} + \frac{1}{n} + \ldots + \frac{1}{n}}$$   Product of Powers

$$= b^1 \text{ or } b$$   Simplify.

Thus, $b^{\frac{1}{n}}$ is a number with an $n$th power equal to $b$. So $b^{\frac{1}{n}} = \sqrt[n]{b}$.

**KeyConcept** $b^{\frac{1}{n}}$

| | |
|---|---|
| **Words** | For any positive real number $b$ and any integer $n > 1$, $b^{\frac{1}{n}} = \sqrt[n]{b}$. |
| **Example** | $8^{\frac{1}{3}} = \sqrt[3]{8} = \sqrt[3]{2 \cdot 2 \cdot 2}$ or 2 |

### Example 3 Evaluate $b^{\frac{1}{n}}$ Expressions

**StudyTip**

**Rational Exponents on a Calculator** Use parentheses to evaluate expressions involving rational exponents on a graphing calculator. For example to find $125^{\frac{1}{3}}$, press 125  ENTER.

**Simplify.**

**a.** $125^{\frac{1}{3}}$

$$125^{\frac{1}{3}} = \sqrt[3]{125} \qquad b^{\frac{1}{n}} = \sqrt[n]{b}$$

$$= \sqrt[3]{5 \cdot 5 \cdot 5} \quad 125 = 5^3$$

$$= 5 \qquad \text{Simplify.}$$

**b.** $1296^{\frac{1}{4}}$

$$1296^{\frac{1}{4}} = \sqrt[4]{1296} \qquad b^{\frac{1}{n}} = \sqrt[n]{b}$$

$$= \sqrt[4]{6 \cdot 6 \cdot 6 \cdot 6} \quad 1296 = 6^4$$

$$= 6 \qquad \text{Simplify.}$$

▶ **GuidedPractice**

**3A.** $27^{\frac{1}{3}}$

**3B.** $256^{\frac{1}{4}}$

The Power of a Power property allows us to extend the definition of $b^{\frac{1}{n}}$ to $b^{\frac{m}{n}}$.

$$b^{\frac{m}{n}} = \left(b^{\frac{1}{n}}\right)^m \qquad \text{Power of a Power}$$

$$= \left(\sqrt[n]{b}\right)^m \text{ or } \sqrt[n]{b^m} \qquad b^{\frac{1}{n}} = \sqrt[n]{b}$$

---

### ⚙ KeyConcept $b^{\frac{m}{n}}$

**Words**     For any positive real number $b$ and any integers $m$ and $n > 1$,
$$b^{\frac{m}{n}} = \left(\sqrt[n]{b}\right)^m \text{ or } \sqrt[n]{b^m}.$$

**Example**     $8^{\frac{2}{3}} = \left(\sqrt[3]{8}\right)^2 = 2^2 \text{ or } 4$

---

### Example 4 Evaluate $b^{\frac{m}{n}}$ Expressions

**Simplify.**

**a.** $64^{\frac{2}{3}}$

$$64^{\frac{2}{3}} = \left(\sqrt[3]{64}\right)^2 \qquad b^{\frac{m}{n}} = \left(\sqrt[n]{b}\right)^m$$

$$= \left(\sqrt[3]{4 \cdot 4 \cdot 4}\right)^2 \quad 64 = 4^3$$

$$= 4^2 \text{ or } 16 \qquad \text{Simplify.}$$

**b.** $36^{\frac{3}{2}}$

$$36^{\frac{3}{2}} = \left(\sqrt[2]{36}\right)^3 \quad b^{\frac{m}{n}} = \left(\sqrt[n]{b}\right)^m$$

$$= 6^3 \qquad \sqrt{36} = 6$$

$$= 216 \qquad \text{Simplify.}$$

▶ **GuidedPractice**

**4A.** $27^{\frac{2}{3}}$

**4B.** $256^{\frac{5}{4}}$

## 2 Solve Exponential Equations

In an **exponential equation**, variables occur as exponents. The Power Property of Equality and the other properties of exponents can be used to solve exponential equations.

> ### 🔑 KeyConcept  Power Property of Equality
>
> **Words**  For any real number $b > 0$ and $b \neq 1$, $b^x = b^y$ if and only if $x = y$.
>
> **Examples**  If $5^x = 5^3$, then $x = 3$. If $n = \frac{1}{2}$, then $4^n = 4^{\frac{1}{2}}$.

### Example 5  Solve Exponential Equations

Solve each equation.

**a.** $6^x = 216$

| | |
|---|---|
| $6^x = 216$ | Original equation |
| $6^x = 6^3$ | Rewrite 216 as $6^3$. |
| $x = 3$ | Property of Equality |

$$\text{CHECK} \quad 6^x = 216$$
$$6^3 \overset{?}{=} 216$$
$$216 = 216 \checkmark$$

**b.** $25^{x-1} = 5$

| | |
|---|---|
| $25^{x-1} = 5$ | Original equation |
| $(5^2)^{x-1} = 5$ | Rewrite 9 as $3^2$. |
| $5^{2x-2} = 5^1$ | Power of a Power, Distributive Property |
| $2x - 2 = 1$ | Power Property of Equality |
| $2x = 3$ | Add 2 to each side. |
| $x = \frac{3}{2}$ | Divide each side by 2. |

$$\text{CHECK} \quad 25^{x-1} = 5$$
$$25^{\frac{3}{2}-1} \overset{?}{=} 5$$
$$25^{\frac{1}{2}} = 5 \checkmark$$

▶ **Guided**Practice

**5A.** $5^x = 125$

**5B.** $12^{2x+3} = 144$

### 🌐 Real-World Example 6  Solve Exponential Equations

**SUNSCREEN** Refer to the beginning of the lesson. Find the SPF that absorbs 100% of UV-B rays.

| | |
|---|---|
| $p = 50f^{0.2}$ | Original equation |
| $100 = 50f^{0.2}$ | $p = 100$ |
| $2 = f^{0.2}$ | Divide each side by 50. |
| $2 = f^{\frac{1}{5}}$ | $0.2 = \frac{1}{5}$ |
| $(2^5)^{\frac{1}{5}} = f^{\frac{1}{5}}$ | $2 = 2^1 = (2^5)^{\frac{1}{5}}$ |
| $2^5 = f$ | Power Property of Equality |
| $32 = f$ | Simplify. |

▶ **Guided**Practice

**6. CHEMISTRY** The radius $r$ of the nucleus of an atom of mass number $A$ is $r = 1.2A^{\frac{1}{3}}$ femtometers. Find $A$ if $r = 3.6$ femtometers.

**Real-World**Link

Use extra caution near snow, water, and sand because they reflect the damaging rays of the Sun, which can increase your chance of sunburn.

**Source:** American Academy of Dermatology

**Example 1**  Write each expression in radical form, or write each radical in exponential form.

**1.** $12^{\frac{1}{2}}$       **2.** $3x^{\frac{1}{2}}$       **3.** $\sqrt{33}$       **4.** $\sqrt{8n}$

**Examples 2–4** Simplify.

**5.** $\sqrt[3]{512}$       **6.** $\sqrt[5]{243}$       **7.** $343^{\frac{1}{3}}$       **8.** $\left(\dfrac{1}{16}\right)^{\frac{1}{4}}$

**9.** $343^{\frac{2}{3}}$       **10.** $81^{\frac{3}{4}}$       **11.** $216^{\frac{4}{3}}$       **12.** $\left(\dfrac{1}{49}\right)^{\frac{3}{2}}$

**Example 5**  Solve each equation.

**13.** $8^x = 4096$       **14.** $3^{3x+1} = 81$       **15.** $4^{x-3} = 32$

**Example 6**  **16. ECOLOGY** A weir is used to measure water flow in a channel. For a rectangular broad crested weir, the flow $Q$ in cubic feet per second is related to the weir length $L$ in feet and height

$H$ of the water by $Q = 1.6LH^{\frac{3}{2}}$. Find the water height for a weir that is 3 feet long and has flow of 38.4 cubic feet per second.

**Example 1**  Write each expression in radical form, or write each radical in exponential form.

**17.** $15^{\frac{1}{2}}$       **18.** $24^{\frac{1}{2}}$       **19.** $4k^{\frac{1}{2}}$       **20.** $(12y)^{\frac{1}{2}}$

**21.** $\sqrt{26}$       **22.** $\sqrt{44}$       **23.** $2\sqrt{ab}$       **24.** $\sqrt{3xyz}$

**Examples 2–4** Simplify.

**25.** $\sqrt[3]{8}$       **26.** $\sqrt[5]{1024}$       **27.** $\sqrt[3]{216}$       **28.** $\sqrt[4]{10{,}000}$

**29.** $\sqrt[3]{0.001}$       **30.** $\sqrt[4]{\dfrac{16}{81}}$       **31.** $1331^{\frac{1}{3}}$       **32.** $64^{\frac{1}{6}}$

**33.** $3375^{\frac{1}{3}}$       **34.** $512^{\frac{1}{9}}$       **35.** $\left(\dfrac{1}{81}\right)^{\frac{1}{4}}$       **36.** $\left(\dfrac{3125}{32}\right)^{\frac{1}{5}}$

**37.** $8^{\frac{2}{3}}$       **38.** $625^{\frac{3}{4}}$       **39.** $729^{\frac{5}{6}}$       **40.** $256^{\frac{3}{8}}$

**41.** $125^{\frac{4}{3}}$       **42.** $49^{\frac{5}{2}}$       **43.** $\left(\dfrac{9}{100}\right)^{\frac{3}{2}}$       **44.** $\left(\dfrac{8}{125}\right)^{\frac{4}{3}}$

**Example 5** **Solve each equation.**

**45.** $3^x = 243$

**46.** $12^x = 144$

**47.** $16^x = 4$

**48.** $27^x = 3$

**49.** $9^x = 27$

**50.** $32^x = 4$

**51.** $2^{x-1} = 128$

**52.** $4^{2x+1} = 1024$

**53.** $6^{x-4} = 1296$

**54.** $9^{2x+3} = 2187$

**55.** $4^{3x} = 512$

**56.** $128^{3x} = 8$

**Example 6** **57. CONSERVATION** Water collected in a rain barrel can be used to water plants and reduce city water use. Water flowing from an open rain barrel has velocity $v = 8h^{\frac{1}{2}}$, where $v$ is in feet per second and $h$ is the height of the water in feet. Find the height of the water if it is flowing at 16 feet per second.

**58. ELECTRICITY** The radius $r$ in millimeters of a platinum wire $L$ centimeters long with resistance 0.1 ohm is $r = 0.059L^{\frac{1}{2}}$. How long is a wire with radius 0.236 millimeters?

**Write each expression in radical form, or write each radical in exponential form.**

**59.** $17^{\frac{1}{3}}$

**60.** $q^{\frac{1}{4}}$

**61.** $7b^{\frac{1}{3}}$

**62.** $m^{\frac{2}{3}}$

**63.** $\sqrt[3]{29}$

**64.** $\sqrt[5]{h}$

**65.** $2\sqrt[3]{a}$

**66.** $\sqrt[3]{xy^2}$

**Simplify.**

**67.** $\sqrt[3]{0.027}$

**68.** $\sqrt[4]{\dfrac{n^4}{16}}$

**69.** $a^{\frac{1}{3}} \cdot a^{\frac{2}{3}}$

**70.** $c^{\frac{1}{2}} \cdot c^{\frac{3}{2}}$

**71.** $(8^2)^{\frac{2}{3}}$

**72.** $\left(y^{\frac{3}{4}}\right)^{\frac{1}{2}}$

**73.** $9^{-\frac{1}{2}}$

**74.** $16^{-\frac{3}{2}}$

**75.** $(3^2)^{-\frac{3}{2}}$

**76.** $\left(81^{\frac{1}{4}}\right)^{-2}$

**77.** $k^{-\frac{1}{2}}$

**78.** $\left(d^{\frac{4}{3}}\right)^0$

**Solve each equation.**

**79.** $2^{5x} = 8^{2x-4}$

**80.** $81^{2x-3} = 9^{x+3}$

**81.** $2^{4x} = 32^{x+1}$

**82.** $16^x = \dfrac{1}{2}$

**83.** $25^x = \dfrac{1}{125}$

**84.** $6^{8-x} = \dfrac{1}{216}$

**85. MUSIC** The frequency $f$ in hertz of the $n$th key on a piano is $f = 440\left(2^{\frac{1}{12}}\right)^{n-49}$.

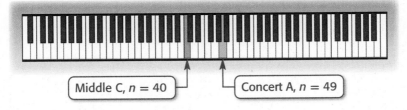

Middle C, $n = 40$      Concert A, $n = 49$

**a.** What is the frequency of Concert A?

**b.** Which note has a frequency of 220 Hz?

86. **RANDOM WALKS** Suppose you go on a walk where you choose the direction of each step at random. The path of a molecule in a liquid or a gas, the path of a foraging animal, and a fluctuating stock price are all modeled as random walks. The number of possible random walks $w$ of $n$ steps where you choose one of $d$ directions at each step is $w = d^n$.

   a. How many steps have been taken in a 2-direction random walk if there are 4096 possible walks?

   b. How many steps have been taken in a 4-direction random walk if there are 65,536 possible walks?

   c. If a walk of 7 steps has 2187 possible walks, how many directions could be taken at each step?

87. **SOCCER** The radius $r$ of a ball that holds $V$ cubic units of air is modeled by $r = 0.62V^{\frac{1}{3}}$. What are the possible volumes of each size soccer ball?

| Soccer Ball Dimensions | |
|---|---|
| Size | Diameter (in.) |
| 3 | 7.3–7.6 |
| 4 | 8.0–8.3 |
| 5 | 8.6–9.0 |

88. 🔁 **MULTIPLE REPRESENTATIONS** In this problem, you will explore the graph of an exponential function.

   a. **TABULAR** Copy and complete the table below.

| $x$ | $-2$ | $-\frac{3}{2}$ | $-1$ | $-\frac{1}{2}$ | $0$ | $\frac{1}{2}$ | $1$ | $\frac{3}{2}$ | $2$ |
|---|---|---|---|---|---|---|---|---|---|
| $f(x) = 4^x$ | | | | | | | | | |

   b. **GRAPHICAL** Graph $f(x)$ by plotting the points and connecting them with a smooth curve.

   c. **VERBAL** Describe the shape of the graph of $f(x)$. What are its key features? Is it linear?

89. **OPEN ENDED** Write two different expressions with rational exponents equal to $\sqrt{2}$.

90. **REASONING** Determine whether each statement is *always*, *sometimes*, or *never* true. Assume that $x$ is a nonnegative real number. Explain your reasoning.

   a. $x^2 = x^{\frac{1}{2}}$

   b. $x^{-2} = x^{\frac{1}{2}}$

   c. $x^{\frac{1}{3}} = x^{\frac{1}{2}}$

   d. $\sqrt{x} = x^{\frac{1}{2}}$

   e. $\left(x^{\frac{1}{2}}\right)^2 = x$

   f. $x^{\frac{1}{2}} \cdot x^2 = x$

91. **CHALLENGE** For what values of $x$ is $x = x^{\frac{1}{3}}$?

92. **ERROR ANALYSIS** Anna and Jamal are solving $128^x = 4$. Is either of them correct? Explain your reasoning.

   Anna
   $128^x = 4$
   $(2^7)^x = 2^2$
   $2^{7x} = 2^2$
   $7x = 2$
   $x = \frac{2}{7}$

   Jamal
   $128^x = 4$
   $(2^7)^x = 4$
   $2^{7x} = 4^1$
   $7x = 1$
   $x = \frac{1}{7}$

93. **WRITING IN MATH** Explain why 2 is the principal fourth root of 16.

# Graphing Technology Lab
# Family of Quadratic Functions

You have studied the effects of changing parameters in the equations of linear and exponential functions. You can use a graphing calculator to analyze how changing the parameters of the equation of a quadratic function affects the graphs in the family of quadratic functions.

## Activity 1  Change $k$ in $y = a(x - h)^2 + k$

**Graph the set of equations on the same screen in the standard viewing window. Describe any similarities and differences among the graphs.**

$y = x^2, y = x^2 + 2, y = x^2 - 4$

Enter the equations in the **Y =** list and graph in the standard viewing window. Use the **ZOOM** feature to investigate the key features of the graphs.

The graphs have the same shape, and all open up. The vertex of each graph is on the $y$-axis, which is the axis of symmetry.

However, the graphs have different vertical positions. The graph of $y = x^2 + 2$ is shifted up 2 units. The graph of $y = x^2 - 4$ is shifted down 4 units.

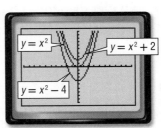

$[-10, 10]$ scl: 1 by $[-10, 10]$ scl: 1

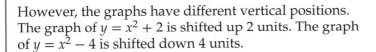

Changing the value of $h$ in $y = a(x - h)^2 + k$ affects the graphs in a different way than changing $k$.

## Activity 2  Change $h$ in $y = a(x - h)^2 + k$

**Graph the set of equations on the same screen in the standard viewing window. Describe any similarities and differences among the graphs.**

$y = x^2, y = (x + 2)^2, y = (x - 4)^2$

The graphs have the same shape, and all open up. The vertex of each graph is on the $x$-axis.

However, the graphs have different horizontal positions. Each has a different axis of symmetry. The graph of $y = (x + 2)^2$ is shifted to the left 2 units. The graph of $y = (x - 4)^2$ is shifted to the right 4 units.

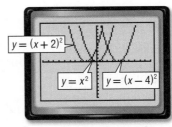

$[-10, 10]$ scl: 1 by $[-10, 10]$ scl: 1

It appears that changing the values of $h$ and $k$ in $y = a(x - h)^2 + k$ moves the graph vertically or horizontally. How does changing the value of $a$ affect the graphs?

*(continued on the next page)*

**Activity 3** Change $a$ in $y = a(x - h)^2 + k$

**Graph each set of equations on the same screen in the standard viewing window. Describe any similarities and differences among the graphs.**

**a.** $y = x^2$, $y = 2x^2$, $y = \frac{1}{3}x^2$

The graphs have the same vertex, they have the same axis of symmetry, and all open up.

However, the graphs have different widths. The graph of $y = 2x^2$ is narrower than the graph of $y = x^2$. The graph of $y = \frac{1}{3}x^2$ is wider than the graph of $y = x^2$.

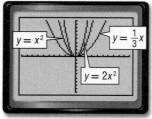

$[-10, 10]$ scl: 1 by $[-10, 10]$ scl: 1

**b.** $y = x^2$, $y = -\frac{1}{3}x^2$, $y = -2x^2$

The graphs have the same vertex and the same axis of symmetry.

However, the graphs of $y = -\frac{1}{3}x^2$ and $y = -2x^2$ open down. Also the graph of $y = -2x^2$ is narrower than the graph of $y = x^2$. The graph of $y = -\frac{1}{3}x^2$ is wider than the graph of $y = x^2$.

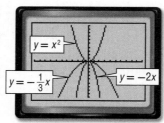

$[-10, 10]$ scl: 1 by $[-10, 10]$ scl: 1

## Model and Analyze

**How does each parameter affect the graph of $y = a(x - h)^2 + k$? Give examples.**

**1.** $k$                                   **2.** $h$                                **3.** $a$

**Examine each pair of equations and predict the similarities and differences in their graphs. Use a graphing calculator to confirm your predictions. Write a sentence or two comparing the two graphs.**

**4.** $y = x^2$, $y = x^2 + 3$                          **5.** $y = \frac{1}{2}x^2$, $y = 3x^2$

**6.** $y = x^2$, $y = (x - 5)^2$                     **7.** $y = 3x^2$, $y = -3x^2$

**8.** $y = x^2$, $y = -4x^2$                           **9.** $y = x^2 - 1$, $y = x^2 + 2$

**10.** $y = \frac{1}{2}x^2 + 3$, $y = -2x^2$             **11.** $y = x^2 - 4$, $y = (x - 4)^2$

# Transformations of Quadratic Functions

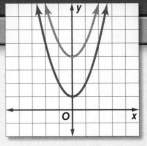

- You graphed quadratic functions by using the vertex and axis of symmetry.

1 Apply translations to quadratic functions.

2 Apply dilations and reflections to quadratic functions.

- The graphs of the parabolas shown at the right are the same size and shape, but notice that the vertex of the red parabola is higher on the $y$-axis than the vertex of the blue parabola. Shifting a parabola up and down is an example of a transformation.

## NewVocabulary
transformation
translation
dilation
reflection
vertex form

**1 Translations** A **transformation** changes the position or size of a figure. One transformation, a **translation**, moves a figure up, down, left, or right. When a constant $k$ is added to or subtracted from the parent function, the graph of the resulting function $f(x) \pm k$ is the graph of the parent function translated up or down.

The parent function of the family of quadratics is $f(x) = x^2$. All other quadratic functions have graphs that are transformations of the graph of $f(x) = x^2$.

### KeyConcept Vertical Translations

The graph of $f(x) = x^2 + k$ is the graph of $f(x) = x^2$ translated vertically.

If $k > 0$, the graph of $f(x) = x^2$ is translated $|k|$ units **up**.

If $k < 0$, the graph of $f(x) = x^2$ is translated $|k|$ units **down**.

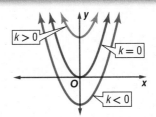

### Example 1 Describe and Graph Translations

**Describe how the graph of each function is related to the graph of $f(x) = x^2$.**

**a.** $h(x) = x^2 + 3$

$k = 3$ and $3 > 0$
$h(x)$ is a translation of the graph of $f(x) = x^2$ up 3 units.

**b.** $g(x) = x^2 - 4$

$k = -4$ and $-4 < 0$
$g(x)$ is a translation of the graph of $f(x) = x^2$ down 4 units.

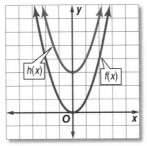

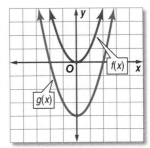

▶ **Guided**Practice

**1A.** $f(x) = x^2 - 7$    **1B.** $g(x) = 5 + x^2$    **1C.** $h(x) = -5 + x^2$    **1D.** $f(x) = x^2 + 1$

A quadratic graph can be translated horizontally by subtracting an *h* term from *x*.

The graph of $g(x) = (x - h)^2$ is the graph of $f(x) = x^2$ translated horizontally.

If $h > 0$, the graph of $f(x) = x^2$ is translated *h* units to the **right**.

If $h < 0$, the graph of $f(x) = x^2$ is translated $|h|$ units to the **left**.

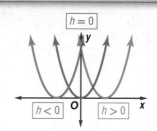

**Example 2** Horizontal Translations

**Describe how the graph of each function is related to the graph of $f(x) = x^2$.**

**a.** $g(x) = (x - 2)^2$

$k = 0, h = 2$ and $2 > 0$
$g(x)$ is a translation of the graph of $f(x) = x^2$ to the right 2 units.

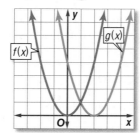

**b.** $g(x) = (x + 1)^2$

$k = 0, h = -1$ and $-1 < 0$
$g(x)$ is a translation of the graph of $f(x) = x^2$ to the left 1 unit.

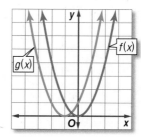

▶ **Guided**Practice

**2A.** $g(x) = (x - 3)^2$

**2B.** $g(x) = (x + 2)^2$

A quadratic graph can be translated both horizontally and vertically.

**Example 3** Horizontal and Vertical Translations

**Describe how the graph of each function is related to the graph of $f(x) = x^2$.**

**a.** $g(x) = (x - 3)^2 + 2$

$k = 2, h = 3$ and $3 > 0$
$g(x)$ is a translation of the graph of $f(x) = x^2$ to the right 3 units and up 2 units.

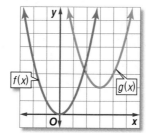

**b.** $g(x) = (x + 3)^2 - 1$

$k = -1, h = -3$ and $-3 < 0$
$g(x)$ is a translation of the graph of $f(x) = x^2$ to the left 3 units and down 1 unit.

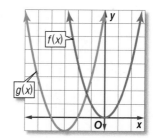

▶ **Guided**Practice

**3A.** $g(x) = (x + 2)^2 + 3$

**3B.** $g(x) = (x - 4)^2 - 4$

## 2 Dilations and Reflections

**2** **Dilations and Reflections** Another type of transformation is a dilation. A **dilation** makes the graph narrower than the parent graph or wider than the parent graph. When the parent function $f(x) = x^2$ is multiplied by a constant $a$, the graph of the resulting function $f(x) = ax^2$ is either stretched or compressed vertically.

---

### KeyConcept Dilations

The graph of $g(x) = ax^2$ is the graph of $f(x) = x^2$ stretched or compressed vertically.

If $|a| > 1$, the graph of $f(x) = x^2$ is stretched vertically.

If $0 < |a| < 1$, the graph of $f(x) = x^2$ is compressed vertically.

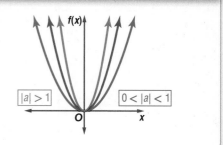

---

**StudyTip**

**Compress or Stretch**
When the graph of a quadratic function is stretched vertically, the shape of the graph is narrower than that of the parent function. When it is compressed vertically, the graph is wider than the parent function.

---

### Example 4 Describe and Graph Dilations

**Describe how the graph of each function is related to the graph of $f(x) = x^2$.**

**a.** $h(x) = \frac{1}{2}x^2$

$a = \frac{1}{2}$ and $0 < \frac{1}{2} < 1$
$h(x)$ is a dilation of the graph of $f(x) = x^2$ that is compressed vertically.

**b.** $g(x) = 3x^2 + 2$

$a = 3$ and $3 > 1$, $k = 2$ and $2 > 0$
$g(x)$ is a dilation of the graph of $f(x) = x^2$ that is stretched vertically and translated up 2 units.

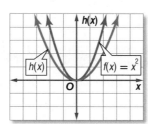   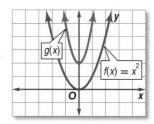

▶ **GuidedPractice**

**4A.** $j(x) = 2x^2$      **4B.** $h(x) = 5x^2 - 2$      **4C.** $g(x) = \frac{1}{3}x^2 + 2$

---

**StudyTip**

**Reflection** A reflection of $f(x) = x^2$ across the $y$-axis results in the same function, because $f(-x) = (-x)^2 = x^2$.

---

A **reflection** flips a figure across a line. When $f(x) = x^2$ or the variable $x$ is multiplied by $-1$, the graph is reflected across the $x$- or $y$-axis.

### KeyConcept Reflections

The graph of $-f(x)$ is the reflection of the graph of $f(x) = x^2$ across the $x$-axis.

The graph of $f(-x)$ is the reflection of the graph of $f(x) = x^2$ across the $y$-axis.

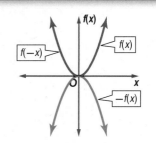

---

**Example 5** Describe and Graph Transformations

Describe how the graph of each function is related to the graph of $f(x) = x^2$.

**a.** $g(x) = -2x^2 - 3$

- $a = -2$, $-2 < 0$, and $|-2| > 1$, so there is a reflection across the $x$-axis and the graph is vertically stretched.
- $k = -3$ and $-3 < 0$, so there is a translation down 3 units.

**b.** $h(x) = -4(x + 2)^2 + 1$

- $a = -4$, $-4 < 0$, and $|-4| > 1$, so there is a reflection across the $x$-axis and the graph is vertically stretched.
- $h = -2$ and $-2 < 0$, so there is a translation 2 units to the left.
- $k = 1$ and $1 > 0$, so there is a translation up 1 unit.

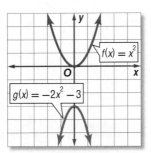

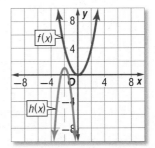

▶ **Guided**Practice

**5A.** $h(x) = 2(-x)^2 - 9$    **5B.** $g(x) = \frac{1}{5}x^2 + 3$    **5C.** $j(x) = -2(x - 1)^2 - 2$

You can use what you know about the characteristics of graphs of quadratic equations to match an equation with a graph.

**Standardized Test Example 6** Identify an Equation for a Graph

Which is an equation for the function shown in the graph?

**A** $y = \frac{1}{2}x^2 - 5$    **C** $y = -\frac{1}{2}x^2 + 5$

**B** $y = -2x^2 - 5$    **D** $y = 2x^2 + 5$

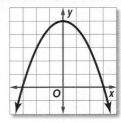

**Read the Test Item**

You are given a graph. You need to find its equation.

**Solve the Test Item**

The graph opens downward, so the graph of $y = x^2$ has been reflected across the $x$-axis. The leading coefficient should be negative, so eliminate choices A and D.

The parabola is translated up 5 units, so $k = 5$. Look at the equations. Only choices C and D have $k = 5$. The answer is C.

▶ **Guided**Practice

**6.** Which is the graph of $y = -3x^2 + 1$?

**F** 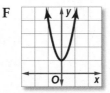    **G** 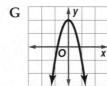    **H** 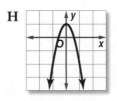    **J**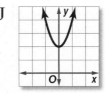

A quadratic function written in the form $f(x) = a(x - h)^2 + k$ is said to be in **vertex form**. Transformations of the parent graph are easily found from an equation in vertex form.

## ConceptSummary Transformations of Quadratic Functions

$$f(x) = a(x - h)^2 + k$$

**$h$, Horizontal Translation**

$h$ units to the right if $h$ is positive

$|h|$ units to the left if $h$ is negative

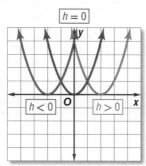

**$k$, Vertical Translation**

$k$ units up if $k$ is positive

$|k|$ units down if $k$ is negative

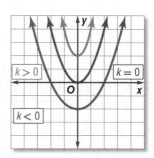

**$a$, Reflection**

If $a > 0$, the graph opens up.

If $a < 0$, the graph opens down.

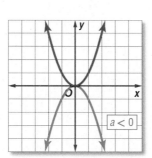

**$a$, Dilation**

If $|a| > 1$, the graph is stretched vertically. If $0 < |a| < 1$, the graph is compressed vertically.

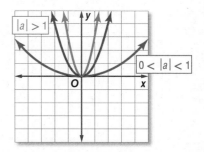

## Real-World Example 7 Transformations with a Calculator

**FIREWORKS** During a firework show, the height $h$ in meters of a specific rocket after $t$ seconds can be modeled by $h(t) = -4.6(t - 3)^2 + 75$. Graph the function. How is it related to the graph of $f(x) = x^2$?

Four separate transformations are occurring.

The negative sign of the coefficient of $x^2$ causes a reflection across the $x$-axis. A dilation occurs, which compresses the graph vertically. There are also translations up 75 units and to the of right 3 units.

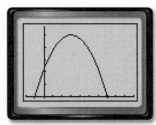

[−2, 10] scl: 1 by [−2, 85] scl: 15

▶ **Guided**Practice

7. **MONUMENTS** The St. Louis Arch resembles a quadratic and can be modeled by $h(x) = -\frac{2}{315}x^2 + 630$. Graph the function. How is it related to the graph of $f(x) = x^2$?

Examples
1–5, 7

Describe how the graph of each function is related to the graph of $f(x) = x^2$.

1. $g(x) = x^2 - 11$

2. $h(x) = \frac{1}{2}(x - 2)^2$

3. $h(x) = -x^2 + 8$

4. $g(x) = x^2 + 6$

5. $g(x) = -4(x + 3)^2$

6. $h(x) = -x^2 - 2$

Example 6

7. **MULTIPLE CHOICE** Which is an equation for the function shown in the graph?

A $g(x) = \frac{1}{5}x^2 + 2$

C $g(x) = \frac{1}{5}x^2 - 2$

B $g(x) = -5x^2 - 2$

D $g(x) = -\frac{1}{5}x^2 - 2$

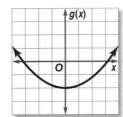

Examples
1–5, 7

Describe how the graph of each function is related to the graph of $f(x) = x^2$.

8. $g(x) = -10 + x^2$

9. $h(x) = -7 - x^2$

10. $g(x) = 2(x - 3)^2 + 8$

11. $h(x) = 6 + \frac{2}{3}x^2$

12. $g(x) = -5 - \frac{4}{3}x^2$

13. $h(x) = 3 + \frac{5}{2}x^2$

14. $g(x) = 0.25x^2 - 1.1$

15. $h(x) = 1.35(x + 1)^2 + 2.6$

16. $g(x) = \frac{3}{4}x^2 + \frac{5}{6}$

17. $h(x) = 1.01x^2 - 6.5$

Example 6

Match each equation to its graph.

A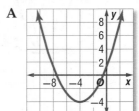

B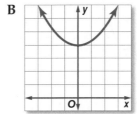

C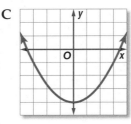

D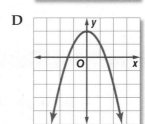

E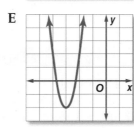

F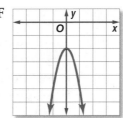

18. $y = \frac{1}{3}x^2 - 4$

19. $y = \frac{1}{3}(x + 4)^2 - 4$

20. $y = \frac{1}{3}x^2 + 4$

21. $y = -3x^2 - 2$

22. $y = -x^2 + 2$

23. $y = (2x + 6)^2 + 2$

24. **SQUIRRELS** A squirrel 12 feet above the ground drops an acorn from a tree. The function $h = -16t^2 + 12$ models the height of the acorn above the ground in feet after $t$ seconds. Graph the function, and compare this graph to the graph of its parent function.

List the functions in order from the most stretched vertically to the least stretched vertically graph.

25. $g(x) = 2x^2, h(x) = \frac{1}{2}x^2$

26. $g(x) = -3x^2, h(x) = \frac{2}{3}x^2$

27. $g(x) = -4x^2, h(x) = 6x^2, f(x) = 0.3x^2$

28. $g(x) = -x^2, h(x) = \frac{5}{3}x^2, f(x) = -4.5x^2$

**29. ROCKS** A rock drops from a cliff 20,000 inches above the ground. At the same time, another rock drops from a cliff 30,000 inches above the ground.

   **a.** Write two functions that model the heights $h$ of the rocks after $t$ seconds.
   **b.** Which rock will reach the ground first?

**30. SPRINKLERS** The path of water from a sprinkler can be modeled by quadratic functions. The following functions model paths for three different sprinklers.

   Sprinkler A: $y = -0.35x^2 + 3.5$       Sprinkler B: $y = -0.21x^2 + 1.7$
   Sprinkler C: $y = -0.08x^2 + 2.4$

   **a.** Which sprinkler will send water the farthest? Explain.
   **b.** Which sprinkler will send water the highest? Explain.
   **c.** Which sprinkler will produce the narrowest path? Explain.

**31. GOLF** The path of a drive can be modeled by a quadratic function where $g(x)$ is the vertical distance in yards of the ball from the ground and $x$ is the horizontal distance in yards.

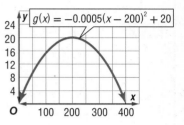

   **a.** How can you obtain $g(x)$ from the graph of $f(x) = x^2$.
   **b.** A second golfer hits a ball from the red tee, which is 30 yards closer to the hole. What function $h(x)$ can be used to describe the second golfer's shot?

**Describe the transformations to obtain the graph of $g(x)$ from the graph of $f(x)$.**

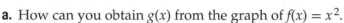

**32.** $f(x) = x^2 + 3$          **33.** $f(x) = x^2 - 4$          **34.** $f(x) = -6x^2$
   $g(x) = x^2 - 2$                 $g(x) = (x - 2)^2 + 7$          $g(x) = -3x^2$

**35. COMBINING FUNCTIONS** An engineer created a self-refueling generator that burns fuel according to the function $g(t) = -t^2 + 10t + 200$, where $t$ represents the time in hours and $g(t)$ represents the number of gallons remaining.

   **a.** How long will it take for the generator to run out of fuel?
   **b.** The engine self-refuels at a rate of 40 gallons per hour. Write a linear function $h(t)$ to represent the refueling of the generator.
   **c.** Find $T(t) = g(t) + h(t)$. What does this new function represent?
   **d.** Will the generator run out of fuel? If so, when?

---

**H.O.T. Problems**  Use Higher-Order Thinking Skills

**36. REASONING** Are the following statements *sometimes*, *always*, or *never* true? Explain.

   **a.** The graph of $y = x^2 + k$ has its vertex at the origin.
   **b.** The graphs of $y = ax^2$ and its reflection over the $x$-axis are the same width.
   **c.** The graph of $y = x^2 + k$, where $k \geq 0$, and the graph of a quadratic with vertex at $(0, -3)$ have the same maximum or minimum point.

**37. CHALLENGE** Write a function of the form $y = ax^2 + k$ with a graph that passes through the points $(-2, 3)$ and $(4, 15)$.

**38. REASONING** Determine whether all quadratic functions that are reflected across the $y$-axis produce the same graph. Explain your answer.

**39. OPEN ENDED** Write a quadratic function that opens downward and is wider than the parent graph.

**40. WRITING IN MATH** Describe how the values of $a$ and $k$ affect the graphical and tabular representations for the functions $y = ax^2$, $y = x^2 + k$, and $y = ax^2 + k$.

In Lesson 9-3, we learned about the vertex form of the equation of a quadratic function. You will now learn how to write equations in vertex form and use them to identify key characteristics of the graphs of quadratic functions.

## Activity 1  Find a Minimum

**Write $y = x^2 + 4x - 10$ in vertex form. Identify the axis of symmetry, extrema, and zeros. Then graph the function.**

**Step 1** Complete the square to write the function in vertex form.

| | |
|---|---|
| $y = x^2 + 4x - 10$ | Original function |
| $y + 10 = x^2 + 4x$ | Add 10 to each side. |
| $y + 10 + 4 = x^2 + 4x + 4$ | Since $\left(\frac{4}{2}\right)^2 = 4$, add 4 to each side. |
| $y + 14 = (x + 2)^2$ | Factor $x^2 + 4x + 4$. |
| $y = (x + 2)^2 - 14$ | Subtract 14 from each side to write in vertex form. |

**Step 2** Identify the axis of symmetry and extrema based on the equation in vertex form. The vertex is at $(h, k)$ or $(-2, -14)$. Since there is no negative sign before the $x^2$-term, the parabola opens up and has a minimum at $(-2, -14)$. The equation of the axis of symmetry is $x = -2$.

**Step 3** Solve for $x$ to find the zeros.

| | |
|---|---|
| $(x + 2)^2 - 14 = 0$ | Vertex form, $y = 0$ |
| $(x + 2)^2 = 14$ | Add 14 to each side. |
| $x + 2 = \pm\sqrt{14}$ | Take square root of each side. |
| $x \approx -5.74 \text{ or } 1.74$ | Subtract 2 from each side. |

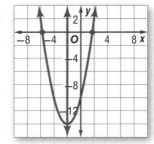

The zeros are approximately $-5.74$ and $1.74$.

**Step 4** Use the key features to graph the function.

There may be a negative coefficient before the quadratic term. When this is the case, the parabola will open down and have a maximum.

## Activity 2  Find a Maximum

**Write $y = -x^2 + 6x - 5$ in vertex form. Identify the axis of symmetry, extrema, and zeros. Then graph the function.**

**Step 1** Complete the square to write the equation of the function in vertex form.

| | |
|---|---|
| $y = -x^2 + 6x - 5$ | Original function |
| $y + 5 = -x^2 + 6x$ | Add 5 to each side. |
| $y + 5 = -(x^2 - 6x)$ | Factor out $-1$. |
| $y + 5 - 9 = -(x^2 - 6x + 9)$ | Since $\left(\frac{6}{2}\right)^2 = 9$, add $-9$ to each side. |
| $y - 4 = -(x - 3)^2$ | Factor $x^2 - 6x + 9$. |
| $y = -(x - 3)^2 + 4$ | Add 4 to each side to write in vertex form. |

**Step 2** Identify the axis of symmetry and extrema based on the equation in vertex form. The vertex is at $(h, k)$ or $(3, 4)$. Since there is a negative sign before the $x^2$-term, the parabola opens down and has a maximum at $(3, 4)$. The equation of the axis of symmetry is $x = 3$.

**Step 3** Solve for $x$ to find the zeros.

$$0 = -(x - 3)^2 + 4 \qquad \text{Vertex form, } y = 0$$
$$(x - 3)^2 = 4 \qquad \text{Add } (x - 3)^2 \text{ to each side.}$$
$$x - 3 = \pm 2 \qquad \text{Take the square root of each side.}$$
$$x = 5 \text{ or } 1 \qquad \text{Add 3 to each.}$$

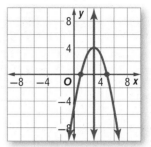

**Step 4** Use the key features to graph the function.

## Analyze the Results

1. Why do you need to complete the square to write the equation of a quadratic function in vertex form?

**Write each function in vertex form. Identify the axis of symmetry, extrema, and zeros. Then graph the function.**

2. $y = x^2 + 6x$     3. $y = x^2 - 8x + 6$     4. $y = x^2 + 2x - 12$

5. $y = x^2 + 6x + 8$     6. $y = x^2 - 4x + 3$     7. $y = x^2 - 2.4x - 2.2$

8. $y = -4x^2 + 16x - 11$     9. $y = 3x^2 - 12x + 5$     10. $y = -x^2 + 6x - 5$

### Activity 3   Use Extrema in the Real World

**DIVING** Alexis jumps from a diving platform upward and outward before diving into the pool. The function $h = -9.8t^2 + 4.9t + 10$, where $h$ is the height of the diver in meters above the pool after $t$ seconds approximates Alexis's dive. Graph the function, then find the maximum height that she reaches and the equation of the axis of symmetry.

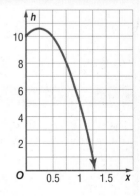

**Step 1** Graph the function.

**Step 2** Complete the square to write the eqution of the function in vertex form.
$$h = -9.8t^2 + 4.9t + 10$$
$$h = -9.8(t - 0.25)^2 + 10.6125$$

**Step 3** The vertex is at $(0.25, 10.6125)$, so the maximum height is 10.6125 meters. The equation of the axis of symmetry is $x = 0.25$.

### Exercise

11. **SOFTBALL** Jenna throws a ball in the air. The function $h = -16t^2 + 40t + 5$, where $h$ is the height in feet and $t$ represents the time in seconds, approximates Jenna's throw. Graph the function, then find the maximum height of the ball and the equation of the axis of symmetry. When does the ball hit the ground?

# Graphing Technology Lab
# Family of Exponential Functions

An **exponential function** is a function of the form $y = ab^x$, where $a \neq 0$, $b > 0$, and $b \neq 1$. You have studied the effects of changing parameters in linear functions. You can use a graphing calculator to analyze how changing the parameters $a$ and $b$ affects the graphs in the family of exponential functions.

## Activity 1   $b$ in $y = b^x$, $b > 1$

**Graph the set of equations on the same screen.**
**Describe any similarities and differences among the graphs.**

$y = 2^x$, $y = 3^x$, $y = 6^x$

Enter the equations in the $\boxed{Y=}$ list and graph.

There are many similarities in the graphs. The domain for each function is all real numbers, and the range is all positive real numbers. The functions are increasing over the entire domain. The graphs do not display any line symmetry.

Use the $\boxed{\text{ZOOM}}$ feature to investigate the key features of the graphs.

Zooming in twice on a point near the origin allows closer inspection of the graphs. The $y$-intercept is 1 for all three graphs.

Tracing along the graphs reveals that there are no $x$-intercepts, no maxima and no minima.

The graphs are different in that the graphs for the equations in which $b$ is greater are steeper.

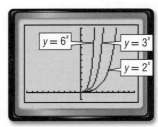

[−10, 10] scl: 1 by [−10, 100] scl: 10

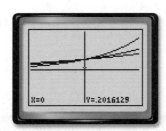

[−0.625, 0.625] scl: 1 by
[−3.25..., 3.63...] scl: 10

The effect of $b$ on the graph is different when $0 < b < 1$.

## Activity 2   $b$ in $y = b^x$, $0 < b < 1$

**Graph the set of equations on the same screen.**
**Describe any similarities and differences among the graphs.**

$y = \left(\frac{1}{2}\right)^x$, $y = \left(\frac{1}{3}\right)^x$, $y = \left(\frac{1}{6}\right)^x$

The domain for each function is all real numbers, and the range is all positive real numbers. The function values are all positive and the functions are decreasing over the entire domain. The graphs display no line symmetry. There are no $x$-intercepts, and the $y$-intercept is 1 for all three graphs. There are no maxima or minima.

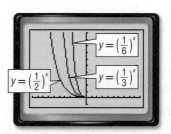

[−10, 10] scl: 1 by [−10, 100] scl: 10

However, the graphs in which $b$ is lesser are steeper.

## Activity 3   $a$ in $y = ab^x$, $a > 0$

**Graph each set of equations on the same screen. Describe any similarities and differences among the graphs.**

$y = 2^x$, $y = 3(2^x)$, $y = \frac{1}{6}(2^x)$

The domain for each function is all real numbers, and the range is all positive real numbers. The functions are increasing over the entire domain. The graphs do not display any line symmetry.

Use the ⬚ZOOM feature to investigate the key features of the graphs.

Zooming in twice on a point near the origin allows closer inspection of the graphs.

Tracing along the graphs reveals that there are no $x$-intercepts, no maxima and no minima.

However, the graphs in which $a$ is greater are steeper. The $y$-intercept is 1 in the graph of $y = 2^x$, 3 in $y = 3(2^x)$, and $\frac{1}{6}$ in $y = \frac{1}{6}(2^x)$.

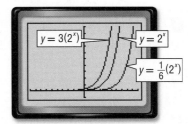

[−10, 10] scl: 1 by [−10, 100] scl: 10

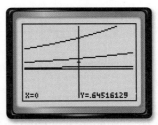

[−0.625, 0.625] scl: 1 by
[−2.79..., 4.08...] scl: 10

## Activity 4   $a$ in $y = ab^x$, $a < 0$

**Graph each set of equations on the same screen. Describe any similarities and differences among the graphs.**

$y = -2^x$, $y = -3(2^x)$, $y = -\frac{1}{6}(2^x)$

The domain for each function is all real numbers, and the range is all negative real numbers. The functions are decreasing over the entire domain. The graphs do not display any line symmetry.

There are no $x$-intercepts, no maxima and no minima.

However, the graphs in which the absolute value of $a$ is greater are steeper. The $y$-intercept is $-1$ in the graph of $y = -2^x$, $-3$ in $y = -3(2^x)$, and $-\frac{1}{6}$ in $y = -\frac{1}{6}(2^x)$.

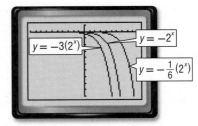

[−10, 10] scl: 1 by [−100, 10] scl: 10

## Model and Analyze

1. How does $b$ affect the graph of $y = ab^x$ when $b > 1$ and when $0 < b < 1$? Give examples.

2. How does $a$ affect the graph of $y = ab^x$ when $a > 0$ and when $a < 0$? Give examples.

3. Make a conjecture about the relationship of the graphs of $y = 3^x$ and $y = \left(\frac{1}{3}\right)^x$. Verify your conjecture by graphing both functions.

# Graphing Technology Lab
# Solving Exponential
# Equations and Inequalities

You can use TI-Nspire Technology to solve exponential equations and inequalities by graphing and by using tables.

---

## Activity 1    Graph an Exponential Equation

**Graph $y = 3^x + 4$ using a graphing calculator.**

**Step 1**  Add a new **Graphs** page.

**Step 2**  Enter $3^x + 4$ as **f1(x)**.

**Step 3**  Use the **Window Settings** option from the **Window/Zoom** menu to adjust the window so that $x$ is from $-10$ to $10$ and $y$ is from $-100$ to $100$. Keep the scales as **Auto**.

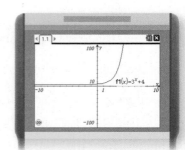

---

To solve an equation by graphing, graph both sides of the equation and locate the point(s) of intersection.

---

## Activity 2    Solve an Exponential Equation by Graphing

**Solve $2^{x-2} = \dfrac{3}{4}$.**

**Step 1**  Add a new **Graphs** page.

**Step 2**  Enter $2^{x-2}$ as **f1(x)** and $\dfrac{3}{4}$ as **f2(x)**.

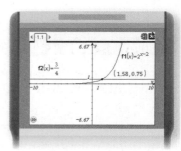

**Step 3**  Use the **Intersection Point(s)** tool from the **Points & Lines** menu to find the intersection of the two graphs. Select the graph of **f1(x) enter** and then the graph of **f2(x) enter**.

The graphs intersect at about $(1.58, 0.75)$. Therefore, the solution of $2^{x-2} = \dfrac{3}{4}$ is 1.58.

---

## Exercises

**Use a graphing calculator to solve each equation.**

**1.** $\left(\dfrac{1}{3}\right)^{x-1} = \dfrac{3}{4}$

**2.** $2^{2x-1} = 2x$

**3.** $\left(\dfrac{1}{2}\right)^{2x} = 2^{2x}$

**4.** $5^{\frac{1}{3}x+2} = -x$

**5.** $\left(\dfrac{1}{8}\right)^{2x} = -2x + 1$

**6.** $2^{\frac{1}{4}x-1} = 3^{x+1}$

**7.** $2^{3x-1} = 4^x$

**8.** $4^{2x-3} = 5^{-x+1}$

**9.** $3^{2x-4} = 2^x + 1$

## Activity 3  Solve an Exponential Equation by Using a Table

**Solve** $2\left(\dfrac{1}{2}\right)^{x+2} = \dfrac{1}{4}$ **using a table.**

**Step 1**  Add a new **Lists & Spreadsheet** page.

**Step 2**  Label column A as $x$. Enter values from $-4$ to $4$ in cells A1 to A9.

**Step 3**  In column B in the formula row, enter the left side of the rational equation. In column C of the formula row, enter $= \dfrac{1}{4}$. Specify **Variable Reference** when prompted.

Scroll until you see where the values in Columns B and C are equal. This occurs at $x = 1$. Therefore, the solution of $2\left(\dfrac{1}{2}\right)^{x+2} = \dfrac{1}{4}$ is 1.

You can also use a graphing calculator to solve exponential inequalities.

## Activity 4  Solve an Exponential Inequality

**Solve** $4^{x-3} \le \left(\dfrac{1}{4}\right)^{2x}$.

**Step 1**  Add a new **Graphs** page.

**Step 2**  Enter the left side of the inequality into **f1(x)**. Press **del**, select $\ge$, and enter $4^{x-3}$. Enter the right side of the inequality into **f2(x)**. Press **tab del** $\le$, and enter $\left(\dfrac{1}{4}\right)^{2x}$.

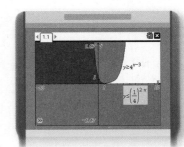

The $x$-values of the points in the region where the shading overlap is the solution set of the original inequality. Therefore the solution of $4^{x-3} \le \left(\dfrac{1}{4}\right)^{2x}$ is $x \le 1$.

## Exercises

Use a graphing calculator to solve each equation or inequality.

**10.** $\left(\dfrac{1}{3}\right)^{3x} = 3^x$

**11.** $\left(\dfrac{1}{6}\right)^{2x} = 4^x$

**12.** $3^{1-x} \le 4^x$

**13.** $4^{3x} \le 2x + 1$

**14.** $\left(\dfrac{1}{4}\right)^{x} > 2^{x+4}$

**15.** $\left(\dfrac{1}{3}\right)^{x-1} \ge 2^x$

You can use the properties of rational exponents to transform exponential functions into other forms in order to solve real-world problems.

### Activity  Write Equivalent Exponential Expressions

**Monique is trying to decide between two savings account plans. Plan A offers 0.25% interest compounded monthly, while Plan B offers 2.5% interest compounded annually. Which is the better plan? Explain.**

In order to compare the plans, we must compare rates with the same compounding frequency. One way to do this is to compare the approximate monthly interest rates of each plan, also called the *effective* monthly interest rate. While you can use the compound interest formula to find this rate, you can also use the properties of exponents.

Write a function to represent the amount $A$ Monique would earn after $t$ years with Plan B. For convenience, let the initial amount of Monique's investment be $1.

$y = a(1 + r)^t$          Equation for exponential growth

$A(t) = 1(1 + 0.025)^t$      $y = A(t)$, $a = 1$, $r = 2.5\%$ or 0.025

$\quad\quad = 1.025^t$                Simplify.

Now write a function equivalent to $A(t)$ that represents 12 compoundings per year, with a power of $12t$, instead of 1 per year, with a power of $1t$.

$A(t) = 1.025^{1t}$         Original function

$\quad\quad = 1.025^{\left(\frac{1}{12} \cdot 12\right)t}$      $1 = \frac{1}{12} \cdot 12$

$\quad\quad = \left(1.025^{\frac{1}{12}}\right)^{12t}$      Power of a Power

$\quad\quad \approx 1.0021^{12t}$      $(1.025)^{\frac{1}{12}} = \sqrt[12]{1.025}$ or about 1.0021

From this equivalent function, we can determine that the effective monthly interest by Plan B is about 0.0021 or about 0.21% per month. This rate is less than the monthly interest rate of 0.25% per month offered by Plan A, so Plan A is the better plan.

### Model and Analyze

1. Use the compound interest formula $A = P\left(1 + \frac{r}{n}\right)^m$ to determine the effective monthly interest rate for Plan B. How does this rate compare to the rate calculated using the method in the Activity above?

2. Write a function to represent the amount $A$ Monique would earn after $t$ months by Plan A. Then use the properties of exponents to write a function equivalent to $A(t)$ that represents the amount earned after $t$ years.

3. From the expression you wrote in Exercise 2, identify the effective annual interest rate by Plan A. Use this rate to explain why Plan A is the better plan.

4. Suppose Plan A offered 1.5% interest per quarter. Use the properties of exponents to explain which is the better plan.

# 15 Algebra Lab
## Average Rate of Change of Exponential Functions

You know that the rate of change of a linear function is the same for any two points on the graph. The rate of change of an exponential function is not constant.

### Activity  Evaluating Investment Plans

John has \$2000 to invest in one of two plans. Plan 1 offers to increase his principal by \$75 each year, while Plan 2 offers to pay 3.6% interest compounded monthly. The dollar value of each investment after $t$ years is given by $A_1 = 2000 + 75t$ and $A_2 = 2000(1.003)^{12t}$, respectively. Use the function values, the average rate of change, and the graphs of the equations to interpret and compare the plans.

**Step 1**  Copy and complete the table below by finding the missing values for $A_1$ and $A_2$.

| $t$ | 0 | 1 | 2 | 3 | 4 | 5 |
|-----|---|---|---|---|---|---|
| $A_1$ | | | | | | |
| $A_2$ | | | | | | |

**Step 2**  Find the average rate of change for each plan from $t = 0$ to 1, $t = 3$ to 4, and $t = 0$ to 5.

Plan 1: $\dfrac{2075 - 2000}{1 - 0}$ or 75 $\qquad$ $\dfrac{2300 - 2225}{4 - 3}$ or 75 $\qquad$ $\dfrac{2375 - 2000}{5 - 0}$ or 75

Plan 2: $\dfrac{2073.2 - 2000}{1 - 0}$ or 73.2 $\qquad$ $\dfrac{2309.27 - 2227.74}{4 - 3}$ or about 82 $\qquad$ $\dfrac{2393.79 - 2000}{5 - 0}$ or about 79

**Step 3**  Graph the ordered pairs for each function. Connect each set of points with a smooth curve.

**Step 4**  Use the graph and the rates of change to compare the plans. Both graphs have a rate of change for the first year of about \$75 per year. From year 3 to 4, Plan 1 continues to increase at \$75 per year, but Plan 2 grows at a rate of more than \$81 per year. The average rate of change over the first five years for Plan 1 is \$75 per year and for Plan 2 is over \$78 per year. This indicates that as the number of years increases, the investment in Plan 2 grows at an increasingly faster pace. This is supported by the widening gap between their graphs.

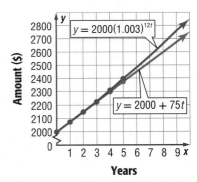

### Exercises

The value of a company's piece of equipment decreases over time due to depreciation. The function $y = 16{,}000(0.985)^{2t}$ represents the value after $t$ years.

1. What is the average rate of change over the first five years?

2. What is the average rate of change of the value from year 5 to year 10?

3. What conclusion about the value can we make based on these average rates of change?

4. Copy and complete the table for $y = x^4$.

| $x$ | −3 | −2 | −1 | 0 | 1 | 2 | 3 |
|-----|----|----|----|---|---|---|---|
| $y$ | | | | | | | |

Compare and interpret the average rate of change for $x = -3$ to 0 and for $x = 0$ to 3.

# Recursive Formulas

| :: Then | :: Now | :: Why? |
|---------|--------|---------|
| • You wrote explicit formulas to represent arithmetic and geometric sequences. | **1** Use a recursive formula to list terms in a sequence. <br><br> **2** Write recursive formulas for arithmetic and geometric sequences. | • Clients of a shuttle service get picked up from their homes and driven to premium outlet stores for shopping. The total cost of the service depends on the total number of customers. The costs for first six customers are shown. |

| Number of Customers | Cost ($) |
|---------------------|----------|
| 1 | 25 |
| 2 | 35 |
| 3 | 45 |
| 4 | 55 |
| 5 | 65 |
| 6 | 75 |

**NewVocabulary**
recursive formula

**1 Using Recursive Formulas** An explicit formula allows you to find any term $a_n$ of a sequence by using a formula written in terms of $n$. For example, $a_n = 2n$ can be used to find the fifth term of the sequence 2, 4, 6, 8, …: $a_5 = 2(5)$ or 10.

A **recursive formula** allows you to find the $n$th term of a sequence by performing operations to one or more of the preceding terms. Since each term in the sequence above is 2 greater than the term that preceded it, we can add 2 to the fourth term to find that the fifth term is $8 + 2$ or 10. We can then write a recursive formula for $a_n$.

$$a_1 = \qquad\qquad\qquad = 2$$
$$a_2 = \quad a_1 + 2 \text{ or } 2 + 2 \quad = 4$$
$$a_3 = \quad a_2 + 2 \text{ or } 4 + 2 \quad = 6$$
$$a_4 = \quad a_3 + 2 \text{ or } 6 + 2 \quad = 8$$
$$\qquad\vdots \qquad\qquad \vdots$$
$$a_n = \qquad a_{n-1} + 2$$

A recursive formula for the sequence above is $a_1 = 2$, $a_n = a_{n-1} + 2$, for $n \geq 2$ where $n$ is an integer. The term denoted $a_{n-1}$ represents the term immediately before $a_n$. Notice that the first term $a_1$ is given, along with the domain for $n$.

**Example 1** Use a Recursive Formula

Find the first five terms of the sequence in which $a_1 = 7$ and $a_n = 3a_{n-1} - 12$, if $n \geq 2$.

Use $a_1 = 7$ and the recursive formula to find the next four terms.

| | | | |
|---|---|---|---|
| $a_2 = 3a_{2-1} - 12$ | $n = 2$ | $a_4 = 3a_{4-1} - 12$ | $n = 4$ |
| $\quad = 3a_1 - 12$ | Simplify. | $\quad = 3a_3 - 12$ | Simplify. |
| $\quad = 3(7) - 12$ or 9 | $a_1 = 7$ | $\quad = 3(15) - 12$ or 33 | $a_3 = 15$ |
| | | | |
| $a_3 = 3a_{3-1} - 12$ | $n = 3$ | $a_5 = 3a_{5-1} - 12$ | $n = 5$ |
| $\quad = 3a_2 - 12$ | Simplify. | $\quad = 3a_4 - 12$ | Simplify. |
| $\quad = 3(9) - 12$ or 15 | $a_2 = 9$ | $\quad = 3(33) - 12$ or 87 | $a_4 = 33$ |

The first five terms are 7, 9, 15, 33, and 87.

▶ **Guided**Practice

**1.** Find the first five terms of the sequence in which $a_1 = -2$ and $a_n = (-3)\,a_{n-1} + 4$, if $n \geq 2$.

# 2 Writing Recursive Formulas
To write a recursive formula for an arithmetic or geometric sequence, complete the following steps.

> **KeyConcept** Writing Recursive Formulas
>
> **Step 1** Determine if the sequence is arithmetic or geometric by finding a common difference or a common ratio.
>
> **Step 2** Write a recursive formula.
>
> Arithmetic Sequences $a_n = a_{n-1} + d$, where $d$ is the common difference
>
> Geometric Sequences $a_n = r \cdot a_{n-1}$, where $r$ is the common ratio
>
> **Step 3** State the first term and domain for $n$.

### Example 2 Write Recursive Formulas

**Write a recursive formula for each sequence.**

**a.** 17, 13, 9, 5, …

> **Step 1** First subtract each term from the term that follows it.
>
> $13 - 17 = -4 \qquad 9 - 13 = -4 \qquad 5 - 9 = -4$
>
> There is a common difference of $-4$. The sequence is arithmetic.

> **Step 2** Use the formula for an arithmetic sequence.
>
> $a_n = a_{n-1} + d$     Recursive formula for arithmetic sequence
>
> $a_n = a_{n-1} + (-4) \quad d = -4$

> **Step 3** The first term $a_1$ is 17, and $n \geq 2$.

A recursive formula for the sequence is $a_1 = 17$, $a_n = a_{n-1} - 4$, $n \geq 2$.

**b.** 6, 24, 96, 384, …

> **Step 1** First subtract each term from the term that follows it.
>
> $24 - 6 = 18 \qquad 96 - 24 = 72 \qquad 384 - 96 = 288$
>
> There is no common difference. Check for a common ratio by dividing each term by the term that precedes it.
>
> $\dfrac{24}{6} = 4 \qquad\qquad \dfrac{96}{24} = 4 \qquad\qquad \dfrac{384}{96} = 4$
>
> There is a common ratio of 4. The sequence is geometric.

> **Step 2** Use the formula for a geometric sequence.
>
> $a_n = r \cdot a_{n-1}$     Recursive formula for geometric sequence
>
> $a_n = 4a_{n-1}$     $r = 4$

> **Step 3** The first term $a_1$ is 6, and $n \geq 2$.

A recursive formula for the sequence is $a_1 = 6$, $a_n = 4a_{n-1}$, $n \geq 2$.

**Guided**Practice

**2A.** 4, 10, 25, 62.5, …

**2B.** 9, 36, 63, 90, …

A sequence can be represented by both an explicit formula and a recursive formula.

---

### Example 3 Write Recursive and Explicit Formulas

**COST** Refer to the beginning of the lesson. Let $n$ be the number of customers.

**a. Write a recursive formula for the sequence.**

**Steps 1 and 2** First subtract each term from the term that follows it.

$$35 - 25 = 10 \qquad 45 - 35 = 10 \qquad 55 - 45 = 10$$

There is a common difference of 10. The sequence is arithmetic.

**Step 3** Use the formula for an arithmetic sequence.

$$a_n = a_{n-1} + d \qquad \text{Recursive formula for arithmetic sequence}$$
$$a_n = a_{n-1} + 10 \qquad d = 10$$

**Step 4** The first term $a_1$ is 25, and $n \geq 2$.

A recursive formula for the sequence is $a_1 = 25, a_n = a_{n-1} + 10, n \geq 2$.

**b. Write an explicit formula for the sequence.**

**Step 1** The common difference is 10.

**Step 2** Use the formula for the $n$th term of an arithmetic sequence.

$$
\begin{aligned}
a_n &= a_1 + (n-1)d \qquad && \text{Formula for the } n\text{th term} \\
&= 25 + (n-1)10 \qquad && a_1 = 25 \text{ and } d = 10 \\
&= 25 + 10n - 10 \qquad && \text{Distributive Property} \\
&= 10n + 15 \qquad && \text{Simplify.}
\end{aligned}
$$

An explicit formula for the sequence is $a_n = 10n + 15$.

▶ **Guided**Practice

3. **SAVINGS** The money that Ronald has in his savings account earns interest each year. He does not make any withdrawals or additional deposits. The account balance at the beginning of each year is $10,000, $10,300, $10,609, $10,927.27, and so on. Write a recursive formula and an explicit formula for the sequence.

---

If several successive terms of a sequence are needed, a recursive formula may be useful, whereas if just the $n$th term of a sequence is needed, an explicit formula may be useful. Thus, it is sometimes beneficial to translate between the two forms.

---

### Example 4 Translate between Recursive and Explicit Formulas

**a. Write a recursive formula for $a_n = 6n + 3$.**

$a_n = 6n + 3$ is an explicit formula for an arithmetic sequence with $d = 6$ and $a_1 = 6(1) + 3$ or 9. Therefore, a recursive formula for $a_n$ is $a_1 = 9, a_n = a_{n-1} + 6, n \geq 2$.

**b. Write an explicit formula for $a_1 = 120, a_n = 0.8a_{n-1}, n \geq 2$.**

$a_n = 0.8a_{n-1}$ is a recursive formula for a geometric sequence with $a_1 = 120$ and $r = 0.8$. Therefore, an explicit formula for $a_n$ is $a_n = 120(0.8)^{n-1}$.

▶ **Guided**Practice

**4A.** Write a recursive formula for $a_n = 4(3)^{n-1}$.

**4B.** Write an explicit formula for $a_1 = -16, a_n = a_{n-1} - 7, n \geq 2$.

**Study**Tip

**Geometric Sequence** Recall that the formula for the $n$th term of a geometric sequence is $a_n = a_1 r^{n-1}$.

**Example 1**  Find the first five terms of each sequence.

**1.** $a_1 = 16, a_n = a_{n-1} - 3, n \geq 2$

**2.** $a_1 = -5, a_n = 4a_{n-1} + 10, n \geq 2$

**Example 2**  Write a recursive formula for each sequence.

**3.** 1, 6, 11, 16, ...

**4.** 4, 12, 36, 108, ...

**Example 3**  **5. BALL**  A ball is dropped from an initial height of 10 feet. The maximum heights the ball reaches on the first three bounces are shown.

**a.** Write a recursive formula for the sequence.

**b.** Write an explicit formula for the sequence.

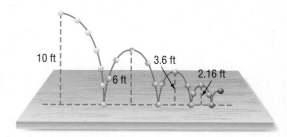

**Example 4**  For each recursive formula, write an explicit formula. For each explicit formula, write a recursive formula.

**6.** $a_1 = 4, a_n = a_{n-1} + 16, n \geq 2$

**7.** $a_n = 5n + 8$

**8.** $a_n = 15(2)^{n-1}$

**9.** $a_1 = 22, a_n = 4a_{n-1}, n \geq 2$

## Practice and Problem Solving

**Example 1**  Find the first five terms of each sequence.

**10.** $a_1 = 23, a_n = a_{n-1} + 7, n \geq 2$

**11.** $a_1 = 48, a_n = -0.5a_{n-1} + 8, n \geq 2$

**12.** $a_1 = 8, a_n = 2.5a_{n-1}, n \geq 2$

**13.** $a_1 = 12, a_n = 3a_{n-1} - 21, n \geq 2$

**14.** $a_1 = 13, a_n = -2a_{n-1} - 3, n \geq 2$

**15.** $a_1 = \frac{1}{2}, a_n = a_{n-1} + \frac{3}{2}, n \geq 2$

**Example 2**  Write a recursive formula for each sequence.

**16.** 12, −1, −14, −27, ...

**17.** 27, 41, 55, 69, ...

**18.** 2, 11, 20, 29, ...

**19.** 100, 80, 64, 51.2, ...

**20.** 40, −60, 90, −135, ...

**21.** 81, 27, 9, 3, ...

**Example 3**  **22. BRICK**  A landscaper is building a brick patio. Part of the patio includes a pattern constructed from triangles. The first four rows of the pattern are shown.

**a.** Write a recursive formula for the sequence.

**b.** Write an explicit formula for the sequence.

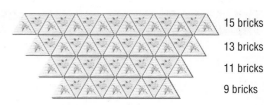

15 bricks
13 bricks
11 bricks
9 bricks

**Example 4**  For each recursive formula, write an explicit formula. For each explicit formula, write a recursive formula.

**23.** $a_n = 3(4)^{n-1}$

**24.** $a_1 = -2, a_n = a_{n-1} - 12, n \geq 2$

**25.** $a_1 = 38, a_n = \frac{1}{2}a_{n-1}, n \geq 2$

**26.** $a_n = -7n + 52$

**27. TEXT MESSAGES** Barbara received a chain text message that she forwarded to five of her friends. Each of her friends forwarded the message to five more friends, and so on.

    **a.** Find the first five terms of the sequence representing the number of people who receive the text in the $n$th round.

    **b.** Write a recursive formula for the sequence.

    **c.** If Barbara represents $a_1$, find $a_8$.

**28. GEOMETRY** Consider the pattern below. The number of blue boxes increases according to a specific pattern.

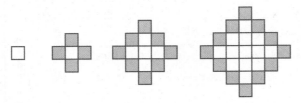

    **a.** Write a recursive formula for the sequence of the number of blue boxes in each figure.

    **b.** If the first box represents $a_1$, find the number of blue boxes in $a_8$.

**29. TREE** The growth of a certain type of tree slows as the tree continues to age. The heights of the tree over the past four years are shown.

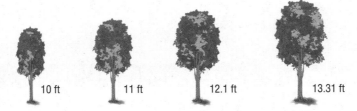

10 ft      11 ft      12.1 ft      13.31 ft

    **a.** Write a recursive formula for the height of the tree.

    **b.** If the pattern continues, how tall will the tree be in two more years? Round your answer to the nearest tenth of a foot.

**30. MULTIPLE REPRESENTATIONS** The Fibonacci sequence is neither arithmetic nor geometric and can be defined by a recursive formula. The first terms are 1, 1, 2, 3, 5, 8, …

    **a. Logical** Determine the relationship between the terms of the sequence. What are the next five terms in the sequence?

    **b. Algebraic** Write a formula for the $n$th term if $a_1 = 1$, $a_2 = 1$, and $n \geq 3$.

    **c. Algebraic** Find the 15th term.

    **d. Analytical** Explain why the Fibonacci sequence is not an arithmetic sequence.

---

**H.O.T. Problems**    Use Higher-Order Thinking Skills

**31. ERROR ANALYSIS** Patrick and Lynda are working on a math problem that involves the sequence 2, −2, 2, −2, 2, … . Patrick thinks that the sequence can be written as a recursive formula. Lynda believes that the sequence can be written as an explicit formula. Is either of them correct? Explain.

**32. CHALLENGE** Find $a_1$ for the sequence in which $a_4 = 1104$ and $a_n = 4a_{n-1} + 16$.

**33. REASONING** Determine whether the following statement is *true* or *false*. Justify your reasoning.

    *There is only one recursive formula for every sequence.*

**34. CHALLENGE** Find a recursive formula for 4, 9, 19, 39, 79, …

**35. WRITING IN MATH** Explain the difference between an explicit formula and a recursive formula.

# Algebra Lab
# Inverse Functions

You have discovered that every nonhorizontal linear function has an inverse function. You have learned how to find the inverse of any function by exchanging the coordinates for a set of ordered pairs. In the following activity, we will exchange coordinates to find the inverse of a quadratic function and determine whether the inverse is a function.

## Activity 1   Exchange Coordinates

**Find the inverse of $y = x^2$ by exchanging the coordinates. Is the inverse a function?**

**Step 1** Make a table of values for $y = x^2$ using $x$ from $-3$ to 3.

| x | −3 | −2 | −1 | 0 | 1 | 2 | 3 |
|---|----|----|----|---|---|---|---|
| y | 9  | 4  | 1  | 0 | 1 | 4 | 9 |

**Step 2** Write the coordinates as a set of ordered pairs.

$\{(-3, 9), (-2, 4), (-1, 1), (0, 0), (1, 1), (2, 4), (3, 9)\}$

**Step 3** Exchange the $x$- and $y$-coordinates in each ordered pair to form the inverse.

$\{(9, -3), (4, -2), \mathbf{(1, -1)}, (0, 0), \mathbf{(1, 1)}, (4, 2), (9, 3)\}$

**Step 4** Examine the set of ordered pairs and determine if it would be a function. This set of ordered pairs would not be a function because each $x$-value is not paired with a unique $y$-value. For example, there are two $y$-values when $x = 1$.

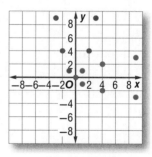

You have also learned how to find the inverse of a linear function algebraically. In the next activity, you will find the inverse of the quadratic function from Activity 1.

## Activity 2   Use Algebra

**Find the inverse of $y = x^2$ algebraically. Check by graphing the function, its inverse, and the line $y = x$.**

**Step 1** Find the inverse algebraically.

$y = x^2$     Original function

$x = y^2$     Interchange $x$ and $y$.

$\pm\sqrt{x} = \sqrt{y^2}$     Take the square root of each side.

$\pm\sqrt{x} = y$     Simplify.

The inverse of $y = x^2$ is $y = \pm\sqrt{x}$.

*(continued on the next page)*

**Step 2** On a coordinate plane, plot and connect the sets of points from Steps 2 and 3 of Activity 1 with a smooth curve to graph $y = x^2$ and its inverse. Graph the line $y = x$.

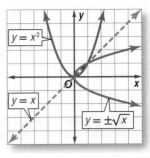

**Step 3** The graph of $y = \pm\sqrt{x}$ does not pass the vertical line test for a function. The inverse is not a function.

Many functions like $y = x^2$ have inverse relations that are not functions. It is often possible to limit the domains of these functions so that their inverses will be functions.

## Activity 3 Restricted Domains

**Restrict the domain of $y = x^2$ so that its inverse is a function.**

Notice from Activity 2 that the graph of $y = x^2$ is symmetric about the $y$-axis. If we restrict the domain of $y = x^2$ to either $x \geq 0$ or $x \leq 0$, we are left with half of the graph.

For $x \geq 0$, the graph of $y = x^2$ is now the portion of the parabola to the right of the $y$-axis. Its inverse is its reflection across the line $y = x$, which is the top portion of the graph of $y = \pm\sqrt{x}$.

Since each $x$-value of this reflection is paired with a unique $y$-value, the inverse is now a function.

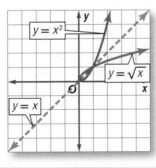

## Exercises

**Write a set of ordered pairs for the inverse of each function by making a table of values for $x$ from $-3$ to $3$ and exchanging the coordinates. Is the inverse a function?**

**1.** $y = x^2 - 3$      **2.** $y = (x - 1)^2$

**3.** $y = 2x^2$      **4.** $y = 3x^2 - 2$

**Find the inverse of each function algebraically. Is the inverse a function?**

**5.** $y = x^2 + 2$      **6.** $y = (x - 1)^2$

**7.** $y = (x + 3)^2 - 4$      **8.** $y = 4x^2 + 2$

**Name a restricted domain for each function for which its inverse would be a function.**

**9.** $y = x^2 - 1$      **10.** $y = (x + 2)^2$

**11.** $y = (x - 2)^2 + 1$      **12.** $y = 3x^2 - 1$

A set is **closed** under an operation if for any numbers in the set, the result of the operation is also in the set. A set may be closed under one operation and not closed under another.

---

### Activity 1  Closure of Rational Numbers and Irrational Numbers

**Are the sets of rational and irrational numbers closed under multiplication? under addition?**

**Step 1** To determine if each set is closed under multiplication, examine several products of two rational factors and then two irrational factors.

Rational: $5 \times 2 = 10$; $-3 \times 4 = -12$; $3.7 \times 0.5 = 1.85$; $\frac{3}{4} \times \frac{2}{3} = \frac{1}{2}$

Irrational: $\pi \times \sqrt{2} = \sqrt{2}\pi$; $\sqrt{3} \times \sqrt{7} = \sqrt{21}$; $\sqrt{5} \times \sqrt{5} = 5$

The product of each pair of rational numbers is rational. However, the products of pairs of irrational numbers are both irrational and rational. Thus, it appears that the set of rational numbers is closed under multiplication, but the set of irrational numbers is not.

**Step 2** Repeat this process for addition.

Rational: $3 + 8 = 11$; $-4 + 7 = 3$; $3.7 + 5.82 = 9.52$; $\frac{2}{5} + \frac{1}{4} = \frac{13}{20}$

Irrational: $\sqrt{3} + \pi = \sqrt{3} + \pi$; $3\sqrt{5} + 6\sqrt{5} = 9\sqrt{5}$; $\sqrt{12} + \sqrt{50} = 2\sqrt{3} + 5\sqrt{2}$

The sum of each pair of rational numbers is rational, and the sum of each pair of irrational numbers is irrational. Both sets are closed under addition.

---

### Activity 2  Rational and Irrational Numbers

**What kind of numbers are the product and sum of a rational and irrational number?**

**Step 1** Examine the products of several pairs of rational and irrational numbers.

$3 \times \sqrt{8} = 6\sqrt{2}$; $\frac{3}{4} \times \sqrt{2} = \frac{3\sqrt{2}}{4}$; $1 \times \sqrt{7} = \sqrt{7}$; $0 \times \sqrt{5} = 0$

The product is rational only when the rational factor is 0. The product of each nonzero rational number and irrational number is irrational.

**Step 2** Find the sums of several pairs of a rational and irrational number.

$5 + \sqrt{3} = 5 + \sqrt{3}$; $\frac{2}{3} + \sqrt{5} = \frac{2 + 3\sqrt{5}}{3}$; $-4 + \sqrt{6} = -1(4 - \sqrt{6})$

The sum of each rational and irrational number is irrational.

---

### Analyze the Results

1. What kinds of numbers are the difference of two unique rational numbers, two unique irrational numbers, and a rational and an irrational number?

2. Is the quotient of every rational and irrational number always another rational or irrational number? If not, provide a counterexample.

3. **CHALLENGE** Recall that rational numbers are numbers that can be written in the form $\frac{a}{b}$, where $a$ and $b$ are integers and $b \neq 0$. Using $\frac{a}{b}$ and $\frac{c}{d}$ show that the sum and product of two rational numbers must always be a rational number.

# Algebra Lab
# Simplifying *n*th
# Root Expressions

The inverse of raising a number to the *n*th power is finding the **nth root** of a number. The **index** of a radical expression indicates to what root the value under the radicand is being taken. The fourth root of a number is indicated with an index of 4. When simplifying a radical expression in which there is a variable with an exponent in the radicand, divide the exponent by the index.

$$13 \div 5 = 2 \text{ R } 3 \longrightarrow \qquad \boxed{\text{index}} \rightarrow \sqrt[5]{x^{13}} = x^2 \cdot \sqrt[5]{x^3} \leftarrow \boxed{\text{remainder}}$$

$$\boxed{\text{quotient}}$$

---

**Example 1** Simplify Expressions

**Simplify each expression.**

a. $\sqrt[3]{x^7}$

$\sqrt[3]{x^7} = x^2 \sqrt[3]{x}$    $7 \div 3 = 2 \text{ R } 1$

b. $\sqrt[5]{32x^9}$

$\sqrt[5]{32x^9} = \sqrt[5]{32} \cdot \sqrt[5]{x^9}$    Multiplication Property

$\qquad\qquad = 2x\sqrt[5]{x^4}$    $9 \div 5 = 1 \text{ R } 4$

---

The properties of square roots (and *n*th roots) also apply when the radicand contains fractions. Separate the numerator and denominator and then simplify them individually.

---

**Example 2** Simplify Expressions with Fractions

**Simplify** $\sqrt[3]{\dfrac{27x^5}{8y^3}}$.

$\sqrt[3]{\dfrac{27x^5}{8y^3}} = \dfrac{\sqrt[3]{27}}{\sqrt[3]{8}} \cdot \dfrac{\sqrt[3]{x^5}}{\sqrt[3]{y^3}}$    Multiplication Property of Radicals

$\qquad\quad = \dfrac{3}{2} \cdot \dfrac{x\sqrt[3]{x^2}}{y}$    Simplify.

$\qquad\quad = \dfrac{3x\sqrt[3]{x^2}}{2y}$    Multiplication Property of Radicals

---

The indices *and* the radicands must be alike in order to add or subtract radical expressions.

---

**Example 3** Combine Like Terms

**Simplify** $8\sqrt[4]{\dfrac{4}{3}} + \sqrt[4]{\dfrac{5}{4}} - 3\sqrt[4]{\dfrac{4}{3}} + \sqrt[3]{\dfrac{4}{3}}$.

Combine the expressions with identical indices and radicands. Then simplify.

$8\sqrt[4]{\dfrac{4}{3}} + \sqrt[4]{\dfrac{5}{4}} - 3\sqrt[4]{\dfrac{4}{3}} + \sqrt[3]{\dfrac{4}{3}} = (8-3)\sqrt[4]{\dfrac{4}{3}} + \sqrt[4]{\dfrac{5}{4}} + \sqrt[3]{\dfrac{4}{3}}$    Associative Property

$\qquad\qquad\qquad\qquad\qquad = 5\sqrt[4]{\dfrac{4}{3}} + \sqrt[4]{\dfrac{5}{4}} + \sqrt[3]{\dfrac{4}{3}}$    Simplify.

---

When multiplying radical expressions, ensure that the indices are the same. Then multiply the radicands and simplify if possible. Once none of the remaining terms can be combined or simplified, the expression is considered simplified.

---

**Example 4** Simplify Expressions with Products

Simplify $5\sqrt[4]{6} \cdot 2\sqrt[4]{12} \cdot \sqrt[3]{10}$.

Multiply the radicands with identical indexes.

$$5\sqrt[4]{6} \cdot 2\sqrt[4]{12} \cdot \sqrt[3]{10} = (5 \cdot 2)(\sqrt[4]{6} \cdot \sqrt[4]{12}) \cdot \sqrt[3]{10} \qquad \text{Associative Property}$$
$$= 10 \cdot (\sqrt[4]{6} \cdot \sqrt[4]{12}) \cdot \sqrt[3]{10} \qquad \text{Multiply.}$$
$$= 10\sqrt[4]{72}\sqrt[3]{10} \qquad \text{Multiply.}$$

---

The properties of radical expressions still hold when variables are in the radicand.

---

**Example 5** Simplify Expressions with Several Operations

Simplify $6\sqrt[4]{x} \cdot \sqrt[4]{x^3} + 3(\sqrt[3]{x} + 2\sqrt[3]{x})$.

Follow the order of operations and the properties of radical expressions.

$$6\sqrt[4]{x} \cdot \sqrt[4]{x^3} + 3(\sqrt[3]{x} + 2\sqrt[3]{x}) = 6\sqrt[4]{x} \cdot \sqrt[4]{x^3} + 3(3\sqrt[3]{x}) \qquad \text{Add like terms.}$$
$$= 6\sqrt[4]{x \cdot x^3} + 3(3\sqrt[3]{x}) \qquad \text{Associative Property}$$
$$= 6\sqrt[4]{x^4} + 9\sqrt[3]{x} \qquad \text{Multiply.}$$
$$= 6x + 9\sqrt[3]{x} \qquad \text{Simplify.}$$

---

## Exercises

Simplify each expression.

1. $\sqrt[3]{c^6}$

2. $\sqrt[4]{16d^9}$

3. $\sqrt[3]{9} \cdot \sqrt[3]{6} \cdot \sqrt[3]{3}$

4. $\sqrt[3]{\dfrac{8a^4}{125b^7}}$

5. $\sqrt[5]{\dfrac{32x^4}{5y^6z^5}}$

6. $\sqrt[4]{\dfrac{3}{2}} + 5\sqrt[4]{\dfrac{3}{2}} - 2\sqrt[4]{\dfrac{2}{3}}$

7. $3\sqrt[4]{6} \cdot 4\sqrt[3]{6} \cdot 5\sqrt[4]{8}$

8. $3\sqrt[4]{x^2} + 2\sqrt[4]{x} \cdot 4\sqrt[4]{x}$

9. $\sqrt[5]{a} \cdot 2\sqrt[5]{a^3} - 2(\sqrt[5]{a} + 4\sqrt[5]{a})$

10. $\sqrt[4]{\dfrac{x}{4}} + 5\sqrt[4]{\dfrac{x}{4}} - 2\sqrt[4]{\dfrac{2x}{3}}$

11. $\sqrt[4]{\dfrac{8a^2}{15b^3}} \cdot 3\sqrt[4]{\dfrac{2a^3}{27b}}$

12. $\sqrt[4]{\dfrac{16x^3}{81y^5}} + 3\sqrt[4]{\dfrac{x^3}{y}} + \sqrt[3]{\dfrac{16x}{y^8}}$

## Think About It

13. Provide an example in which two radical expressions with *unlike* radicands can be combined by addition.

14. Provide an example in which two radical expressions with identical indices and with like variables in the radicand *cannot* be combined by addition.

# Graphing Technology Lab
## Solving Rational Equations

You can use TI-Nspire Technology to solve rational equations by graphing, by using tables, and by using a computer algebra system (CAS).

To solve by graphing, graph both sides of the equation and locate the point(s) of intersection.

---

### Activity 1 Solve a Rational Equation by Graphing

**Solve $\dfrac{5}{x+2} = \dfrac{3}{x}$ by graphing.**

**Step 1** Add a new **Graphs** page.

**Step 2** Use the **Window Settings** option from the **Window/Zoom** menu to adjust the window to −20 to 20 for both $x$ and $y$. Set both scales to 2.

**Step 3** Enter $\dfrac{5}{x+2}$ into **f1(x)** and $\dfrac{3}{x}$ into **f2(x)**.

**Step 4** Change the thickness of the graph of **f1(x)** by selecting the graph of **f1(x)** and the **ctrl menu Attributes** option.

**Step 5** Use the **Intersection Point(s)** tool from the **Points & Lines** menu to find the intersection of the two graphs. Select the graph of **f1(x)** enter and then the graph of **f2(x)** enter.

The graphs intersect at (3, 1). This means that $\dfrac{5}{x+2}$ and $\dfrac{3}{x}$ both equal 1 when $x = 3$. Thus, the solution of $\dfrac{5}{x+2} = \dfrac{3}{x}$ is $x = 3$.

---

### Exercises

**Use a graphing calculator to solve each equation.**

1. $\dfrac{5}{x} + \dfrac{4}{x} = 10$

2. $\dfrac{12}{x} + \dfrac{3}{4} = \dfrac{3}{2}$

3. $\dfrac{6}{x} + \dfrac{3}{2x} = 12$

4. $\dfrac{4}{x} + \dfrac{3}{4x} = \dfrac{1}{8}$

5. $\dfrac{4}{x} + \dfrac{x-2}{2x} = x$

6. $\dfrac{3}{3x-2} + \dfrac{5}{x} = 0$

7. $\dfrac{2x+1}{2} + \dfrac{3}{2x} = \dfrac{2}{x}$

8. $\dfrac{x}{x+2} + x = \dfrac{5x+8}{x+2}$

9. $\dfrac{1}{2x} + \dfrac{5}{x} = \dfrac{3}{x-1}$

10. $\dfrac{4x-3}{x-2} + \dfrac{2x+5}{x-2} = 6$

## Activity 2   Solve a Rational Equation by Using a Table

**Solve $\dfrac{2x+1}{3} = \dfrac{x+2}{2}$ using a table.**

**Step 1** Add a new **Lists & Spreadsheet** page.

**Step 2** Label column A as $x$. Enter values from $-4$ to 4 in cells A1 to A9.

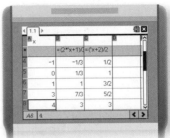

**Step 3** In column B in the formula row, enter the left side of the rational equation, with parenthesis around the binomials. In column C in the formula row, enter the right side of the rational equation, with parenthesis around the binomials. Specify **Variable Reference** when prompted.

Scroll until you see where the values in Columns B and C are equal. This occurs at $x = 4$. Therefore the solution of $\dfrac{2x+1}{3} = \dfrac{x+2}{2}$ is 4.

You can also use a computer algebra system (CAS) to solve rational equations.

## Activity 3   Solve a Rational Equation by Using a CAS

**Solve $\dfrac{x-3}{x} - \dfrac{x-4}{x-2} = \dfrac{1}{x}$ using a CAS.**

**Step 1** Add a new **Calculator** page.

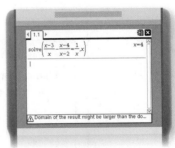

**Step 2** To solve, select the **Solve** tool from the **Algebra** menu. Enter the left side of the equation with parenthesis around the binomials. Enter = and the right side of the equation. Then type a comma, followed by $x$, and then **enter**.

The solution of 4 is displayed.

## Exercises

**Use a table or CAS to solve each equation.**

**11.** $\dfrac{2}{x} + \dfrac{2+x}{2} = \dfrac{x+3}{2}$

**12.** $\dfrac{4}{x-2} = -\dfrac{1}{x+3}$

**13.** $\dfrac{3}{x+2} + \dfrac{4}{x-1} = 0$

**14.** $\dfrac{1}{x+1} + \dfrac{2}{x-1} = 0$

**15.** $\dfrac{2}{x+4} + \dfrac{4}{x-1} = 0$

**16.** $\dfrac{1}{x-2} + \dfrac{x+2}{4} = 2x$

**17.** $\dfrac{2x}{x+3} + \dfrac{x+1}{2} = x$

**18.** $\dfrac{2}{x-3} + \dfrac{3}{x-2} = \dfrac{4}{x}$

**19.** $\dfrac{x^2}{x+1} + \dfrac{x}{x-1} = x$

# Distributions of Data

- You calculated measures of central tendency and variation.

**1** Describe the shape of a distribution.

**2** Use the shapes of distributions to select appropriate statistics.

- While training for the 100-meter dash, Sarah pulled a muscle in her lower back. After being cleared for practice, she continued to train. Sarah's median time was about 12.34 seconds, but her average time dropped to about 12.53 seconds.

 **NewVocabulary**
distribution
negatively skewed
   distribution
symmetric distribution
positively skewed
   distribution

**1** **Describing Distributions** A **distribution** of data shows the observed or theoretical frequency of each possible data value. Recall that a histogram is a type of bar graph used to display data that have been organized into equal intervals. A histogram is useful when viewing the overall distribution of the data within a set over its range. You can see the shape of the distribution by drawing a curve over the histogram.

**KeyConcept** Symmetric and Skewed Distributions

| **Negatively Skewed Distribution** | **Symmetric Distribution** | **Positively Skewed Distribution** |
|---|---|---|
|  | 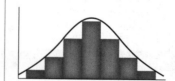 | 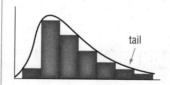 |
| The majority of the data are on the right. | The data are evenly distributed. | The majority of the data are on the left. |

**Example 1** Distribution Using a Histogram

**Use a graphing calculator to construct a histogram for the data, and use it to describe the shape of the distribution.**

25, 22, 31, 25, 26, 35, 18, 39, 22, 32, 34, 26, 42, 23, 40, 36, 18, 30
26, 30, 37, 23, 19, 33, 24, 29, 39, 21, 43, 25, 34, 24, 26, 30, 21, 22

First, press ⎡STAT⎤ ⎡ENTER⎤ and enter each data value.
Then, press ⎡2nd⎤ [STAT PLOT] ⎡ENTER⎤ ⎡ENTER⎤ and choose
⎡▯▯▯⎤. Press ⎡ZOOM⎤ [ZoomStat] to adjust the window.

The graph is high on the left and has a tail on the right. Therefore, the distribution is positively skewed.

[17, 45] scl: 4 by [0, 10] scl: 1

▶ **Guided**Practice

**1.** Use a graphing calculator to construct a histogram for the data, and use it to describe the shape of the distribution.

8, 11, 15, 25, 21, 26, 20, 12, 32, 20, 31, 14, 19, 27, 22, 21, 14, 8
6, 23, 18, 16, 28, 25, 16, 20, 29, 24, 17, 35, 20, 27, 10, 16, 22, 12

A box-and-whisker plot can also be used to identify the shape of a distribution. Recall from Lesson 0-13 that a box-and-whisker plot displays the spread of a data set by dividing it into four quartiles. The data from Example 1 are displayed below.

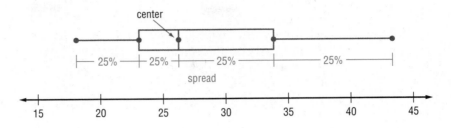

Notice that the left whisker is shorter than the right whisker, and that the line representing the median is closer to the left whisker. This represents a peak on the left and a tail to the right.

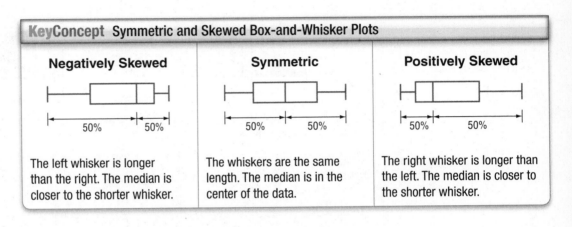

**KeyConcept** Symmetric and Skewed Box-and-Whisker Plots

| **Negatively Skewed** | **Symmetric** | **Positively Skewed** |
|---|---|---|
| The left whisker is longer than the right. The median is closer to the shorter whisker. | The whiskers are the same length. The median is in the center of the data. | The right whisker is longer than the left. The median is closer to the shorter whisker. |

**Example 2** Distribution Using a Box-and-Whisker Plot

**Use a graphing calculator to construct a box-and-whisker plot for the data, and use it to determine the shape of the distribution.**

9, 17, 15, 10, 16, 2, 17, 19, 10, 18, 14, 8, 20, 20, 3, 21, 12, 11
5, 26, 15, 28, 12, 5, 27, 26, 15, 53, 12, 7, 22, 11, 8, 16, 22, 15

Enter the data as **L1**. Press [2nd] [STAT PLOT] [ENTER] [ENTER] and choose ⊡⠇. Adjust the window to the dimensions shown.

The lengths of the whiskers are approximately equal, and the median is in the middle of the data. This indicates that the data are equally distributed to the left and right of the median. Thus, the distribution is symmetric.

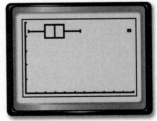

[0, 55] scl: 5 by [0, 5] scl: 1

**StudyTip**

Outliers  In Example 2, notice that the outlier does not affect the shape of the distribution.

**Guided**Practice

**2.** Use a graphing calculator to construct a box-and-whisker plot for the data, and use it to describe the shape of the distribution.

40, 50, 35, 48, 43, 31, 52, 42, 54, 38, 50, 46, 49, 43, 40, 50, 32, 53
51, 43, 47, 41, 49, 50, 34, 54, 51, 44, 54, 39, 47, 35, 51, 44, 48, 37

## 2 Analyzing Distributions

You have learned that data can be described using statistics. The mean and median describe the center. The standard deviation and quartiles describe the spread. You can use the shape of the distribution to choose the most appropriate statistics that describe the center and spread of a set of data.

When a distribution is symmetric, the mean accurately reflects the center of the data. However, when a distribution is skewed, this statistic is not as reliable.

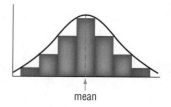

mean

In Lesson 0-12, you discovered that outliers can have a strong effect on the mean of a data set, while the median is less affected. So, when a distribution is skewed, the mean lies away from the majority of the data toward the tail. The median is less affected and stays near the majority of the data.

**Negatively Skewed Distribution**

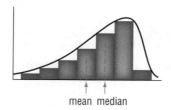

mean median

**Positively Skewed Distribution**

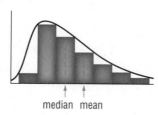

median mean

When choosing appropriate statistics to represent a set of data, first determine the shape of the distribution.

- If the distribution is relatively symmetric, the mean and standard deviation can be used.
- If the distribution is skewed or has outliers, use the five-number summary.

---

### Example 3  Choose Appropriate Statistics

**Describe the center and spread of the data using either the mean and standard deviation or the five-number summary. Justify your choice by constructing a histogram for the data.**

> 21, 28, 16, 30, 25, 34, 21, 47, 18, 36, 24, 28, 30, 15, 33, 24, 32, 22
> 27, 38, 23, 29, 15, 27, 33, 19, 34, 29, 23, 26, 19, 30, 25, 13, 20, 25

Use a graphing calculator to create a histogram. The graph is high in the middle and low on the left and right. Therefore, the distribution is symmetric.

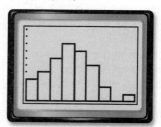

[12, 48] scl: 4 by [0, 10] scl: 1

The distribution is symmetric, so use the mean and standard deviation to describe the center and spread.
Press STAT ▶ ENTER ENTER.

The mean $\bar{x}$ is about 26.1 with standard deviation $\sigma$ of about 7.1.

**3.** Describe the center and spread of the data using either the mean and standard deviation or the five-number summary. Justify your choice by creating a histogram for the data.

> 19, 2, 25, 14, 24, 20, 27, 30, 14, 25, 19, 32, 21, 31, 25, 16, 24, 22
> 29, 6, 26, 32, 17, 26, 24, 26, 32, 10, 28, 19, 26, 24, 11, 23, 19, 8

A box-and-whisker plot is helpful when viewing a skewed distribution since it is constructed using the five-number summary.

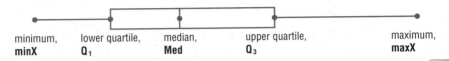

| minimum, **minX** | lower quartile, $Q_1$ | median, **Med** | upper quartile, $Q_3$ | maximum, **maxX** |

### Real-World Example 4  Choose Appropriate Statistics

**COMMUNITY SERVICE** The number of community service hours each of Ms. Tucci's students completed is shown. Describe the center and spread of the data using either the mean and standard deviation or the five-number summary. Justify your choice by constructing a box-and-whisker plot for the data.

| Community Service Hours | | | | | | | | | | | | |
|---|---|---|---|---|---|---|---|---|---|---|---|---|
| 6 | 13 | 8 | 7 | 19 | 12 | 2 | 19 | 11 | 22 | 7 | 33 | 13 |
| 3 | 8 | 10 | 5 | 25 | 16 | 6 | 14 | 7 | 20 | 10 | 30 | |

Use a graphing calculator to create a box-and-whisker plot. The right whisker is longer than the left and the median is closer to the left whisker. Therefore, the distribution is positively skewed.

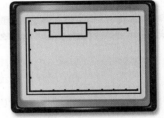

[0, 36] scl: 4 by [0, 5] scl: 1

The distribution is positively skewed, so use the five-number summary. The range is 33 – 2 or 31. The median number of hours completed is 11, and half of the students completed between 7 and 19 hours.

**4. FUNDRAISER** The money raised per student in Mr. Bulanda's 5th period class is shown. Describe the center and spread of the data using either the mean and standard deviation or the five-number summary. Justify your choice by creating a box-and-whisker plot for the data.

| Money Raised per Student (dollars) | | | | | | | | | |
|---|---|---|---|---|---|---|---|---|---|
| 41 | 27 | 52 | 18 | 42 | 32 | 16 | 95 | 27 | 65 |
| 36 | 45 | 5 | 34 | 50 | 15 | 62 | 38 | 57 | 20 |
| 38 | 21 | 33 | 58 | 25 | 42 | 31 | 8 | 40 | 28 |

**Real-WorldLink**

Volunteers in the Peace Corps must be at least 18 years old, and more than 90% of volunteers have college degrees. Volunteers work in another country for 27 months and are placed in host countries that have the greatest needs for skilled volunteers.

**Source:** Peace Corps

**Examples 1–2** Use a graphing calculator to construct a histogram and a box-and-whisker plot for the data. Then describe the shape of the distribution.

1. 80, 84, 68, 64, 57, 88, 61, 72, 76, 80, 83, 77, 78, 82, 65, 70, 83, 78
73, 79, 70, 62, 69, 66, 79, 80, 86, 82, 73, 75, 71, 81, 74, 83, 77, 73

2. 30, 24, 35, 84, 60, 42, 29, 16, 68, 47, 22, 74, 34, 21, 48, 91, 66, 51
33, 29, 18, 31, 54, 75, 23, 45, 25, 32, 57, 40, 23, 32, 47, 67, 62, 23

**Example 3** Describe the center and spread of the data using either the mean and standard deviation or the five-number summary. Justify your choice by constructing a histogram for the data.

3. 58, 66, 52, 75, 60, 56, 78, 63, 59, 54, 60, 67, 72, 80, 68, 88, 55, 60
59, 61, 82, 70, 67, 60, 58, 86, 74, 61, 92, 76, 58, 62, 66, 74, 69, 64

**Example 4** 4. **PRESENTATIONS** The length of the students' presentations in Ms. Monroe's 2nd period class are shown. Describe the center and spread of the data using either the mean and standard deviation or the five-number summary. Justify your choice by constructing a box-and-whisker plot for the data.

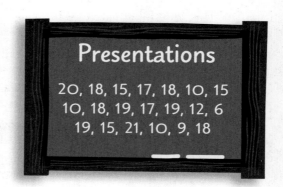

Presentations

20, 18, 15, 17, 18, 10, 15
10, 18, 19, 17, 19, 12, 6
19, 15, 21, 10, 9, 18

**Practice and Problem Solving**

**Examples 1–2** Use a graphing calculator to construct a histogram and a box-and-whisker plot for the data. Then describe the shape of the distribution.

5. 55, 65, 70, 73, 25, 36, 33, 47, 52, 54, 55, 60, 45, 39, 48, 55, 46, 38
50, 54, 63, 31, 49, 54, 68, 35, 27, 45, 53, 62, 47, 41, 50, 76, 67, 49

6. 42, 48, 51, 39, 47, 50, 48, 51, 54, 46, 49, 36, 50, 55, 51, 43, 46, 37
50, 52, 43, 40, 33, 51, 45, 53, 44, 40, 52, 54, 48, 51, 47, 43, 50, 46

**Example 3** Describe the center and spread of the data using either the mean and standard deviation or the five-number summary. Justify your choice by constructing a histogram for the data.

7. 32, 44, 50, 49, 21, 12, 27, 41, 48, 30, 50, 23, 37, 16, 49, 53, 33, 25
35, 40, 48, 39, 50, 24, 15, 29, 37, 50, 36, 43, 49, 44, 46, 27, 42, 47

8. 82, 86, 74, 90, 70, 81, 89, 88, 75, 72, 69, 91, 96, 82, 80, 78, 74, 94
85, 77, 80, 67, 76, 84, 80, 83, 88, 92, 87, 79, 84, 96, 85, 73, 82, 83

**Example 4** 9. **WEATHER** The daily low temperatures for New Carlisle over a 30-day period are shown. Describe the center and spread of the data using either the mean and standard deviation or the five-number summary. Justify your choice by constructing a box-and-whisker plot for the data.

| Temperature (°F) | | | | | | | | | | | | | | |
|---|---|---|---|---|---|---|---|---|---|---|---|---|---|---|
| 48 | 50 | 55 | 53 | 57 | 53 | 44 | 61 | 57 | 49 | 51 | 58 | 46 | 54 | 57 |
| 50 | 55 | 47 | 57 | 48 | 58 | 53 | 49 | 56 | 59 | 52 | 48 | 55 | 53 | 51 |

10. **TRACK** Refer to the beginning of the lesson. Sarah's 100-meter dash times are shown.

   **a.** Use a graphing calculator to create a box-and-whisker plot. Describe the center and spread of the data.

   **b.** Sarah's slowest time prior to pulling a muscle was 12.50 seconds. Use a graphing calculator to create a box-and-whisker plot that *does not* include the times that she ran after pulling the muscle. Then describe the center and spread of the new data set.

   **c.** What effect does removing the times recorded after Sarah pulled a muscle have on the shape of the distribution and on how you should describe the center and spread?

| 100-meter dash (seconds) | | | | |
|---|---|---|---|---|
| 12.20 | 12.35 | 13.60 | 12.24 | 12.72 |
| 12.18 | 12.06 | 12.41 | 12.28 | 13.06 |
| 12.87 | 12.04 | 12.38 | 12.20 | 13.12 |
| 12.30 | 13.27 | 12.93 | 12.16 | 12.02 |
| 12.50 | 12.14 | 11.97 | 12.24 | 13.09 |
| 12.46 | 12.33 | 13.57 | 11.96 | 13.34 |

11. **MENU** The prices for entrees at a restaurant are shown.

   **a.** Use a graphing calculator to create a box-and-whisker plot. Describe the center and spread of the data.

   **b.** The owner of the restaurant decides to eliminate all entrees that cost more than $15. Use a graphing calculator to create a box-and-whisker plot that reflects this change. Then describe the center and spread of the new data set.

| Entree Prices ($) | | | | |
|---|---|---|---|---|
| 9.00 | 11.25 | 16.50 | 9.50 | 13.00 |
| 18.50 | 7.75 | 11.50 | 13.75 | 9.75 |
| 8.00 | 16.50 | 12.50 | 10.25 | 17.75 |
| 13.00 | 10.75 | 16.75 | 8.50 | 11.50 |

## H.O.T. Problems   Use Higher-Order Thinking Skills

**CHALLENGE** Identify the histogram that corresponds to each of the following box-and-whisker plots.

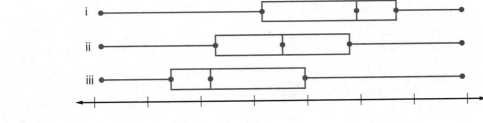

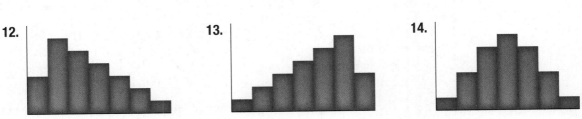

**12.** **13.** **14.**

15. **WRITING IN MATH** Research and write a definition for a *bimodal distribution*. How can the measures of center and spread of a bimodal distribution be described?

16. **OPEN ENDED** Give an example of a set of real-world data with a distribution that is symmetric and one with a distribution that is not symmetric.

17. **WRITING IN MATH** Explain why the mean and standard deviation are used to describe the center and spread of a symmetrical distribution and the five-number summary is used to describe the center and spread of a skewed distribution.

# Comparing Sets of Data

- You calculated measures of central tendency and variation.

**1** Determine the effect that transformations of data have on measures of central tendency and variation.

**2** Compare data using measures of central tendency and variation.

- Tom gets paid hourly to do landscaping work. Because he is such a good employee, Tom is planning to ask his boss for a bonus. Tom's initial pay for a month is shown. He is trying to decide whether he should ask for an extra $5 per day or a 10% increase in his daily wages.

| Tom's Pay ($) | | |
|---|---|---|
| 44 | 52 | 50 |
| 40 | 48 | 46 |
| 44 | 52 | 54 |
| 58 | 42 | 52 |
| 54 | 50 | 52 |
| 42 | 52 | 46 |
| 56 | 48 | 44 |
| 50 | 42 | |

Peter Cade/Iconica/Getty Images

**NewVocabulary**

linear transformation

**1 Transformations of Data** To see the effect that an extra $5 per day would have on Tom's daily pay, we can find the new daily pay values and compare the measures of center and variation for the two sets of data. The new data can be found by performing a *linear transformation*. A **linear transformation** is an operation performed on a data set that can be written as a linear function. Tom's daily pay after the $5 bonus can be found using $y = 5 + x$, where $x$ represents his original daily pay and $y$ represents his daily pay after the bonus.

### Tom's Earnings Before Extra $5

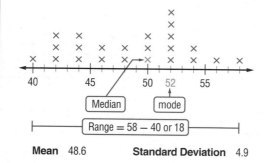

Median 52 mode

Range = 58 − 40 or 18

**Mean** 48.6     **Standard Deviation** 4.9

### Tom's Earnings With Extra $5

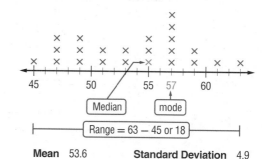

Median 57 mode

Range = 63 − 45 or 18

**Mean** 53.6     **Standard Deviation** 4.9

Notice that each value was translated 5 units to the right. Thus, the mean, median, and mode increased by 5. Since the new minimum and maximum values also increased by 5, the range remained the same. The standard deviation is unchanged because the amount by which each value deviates from the mean stayed the same.

These results occur when any positive or negative number is added to every value in a set of data.

---

**KeyConcept** Transformations Using Addition

If a real number $k$ is added to every value in a set of data, then:

- the mean, median, and mode of the new data set can be found by adding $k$ to the mean, median, and mode of the original data set, and

- the range and standard deviation will not change.

---

**Example 1** Transformation Using Addition

Find the mean, median, mode, range, and standard deviation of the data set obtained after adding 7 to each value.

13, 5, 8, 12, 7, 4, 5, 8, 14, 11, 13, 8

**Method 1** Find the mean, median, mode, range, and standard deviation of the original data set.

| Mean | 9 | Mode | 8 | Standard Deviation | 3.3 |
| Median | 8 | Range | 10 | | |

Add 7 to the mean, median, and mode. The range and standard deviation are unchanged.

| Mean | 16 | Mode | 15 | Standard Deviation | 3.3 |
| Median | 15 | Range | 10 | | |

**Method 2** Add 7 to each data value.

20, 12, 15, 19, 14, 11, 12, 15, 21, 18, 20, 15

Find the mean, median, mode, range, and standard deviation of the new data set.

| Mean | 16 | Mode | 15 | Standard Deviation | 3.3 |
| Median | 15 | Range | 10 | | |

▶ **Guided**Practice

**1.** Find the mean, median, mode, range, and standard deviation of the data set obtained after adding −4 to each value.

27, 41, 15, 36, 26, 40, 53, 38, 37, 24, 45, 26

To see the effect that a daily increase of 10% has on the data set, we can multiply each value by 1.10 and recalculate the measures of center and variation.

**Tom's Earnings Before Extra 10%**

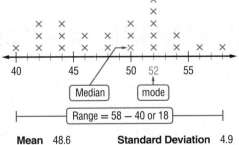

**Mean** 48.6     **Standard Deviation** 4.9

**Tom's Earnings With Extra 10%**

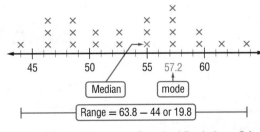

**Mean** 53.5     **Standard Deviation** 5.4

Notice that each value did not increase by the same amount, but did increase by a factor of 1.10. Thus, the mean, median, and mode increased by a factor of 1.10. Since each value was increased by a constant percent and not by a constant amount, the range and standard deviation both changed, also increasing by a factor of 1.10.

**KeyConcept** Transformations Using Multiplication

If every value in a set of data is multiplied by a constant $k$, $k > 0$, then the mean, median, mode, range, and standard deviation of the new data set can be found by multiplying each original statistic by $k$.

Since the medians for both bonuses are equal and the means are approximately equal, Tom should ask for the bonus that he thinks he has the best chance of receiving.

### Example 2 Transformation Using Multiplication

**Find the mean, median, mode, range, and standard deviation of the data set obtained after multiplying each value by 3.**

**21, 12, 15, 18, 16, 10, 12, 19, 17, 18, 12, 22**

Find the mean, median, mode, range, and standard deviation of the original data set.

| Mean | 16 | Mode | 12 | Standard Deviation | 3.7 |
| Median | 16.5 | Range | 12 | | |

Multiply the mean, median, mode, range, and standard deviation by 3.

| Mean | 48 | Mode | 36 | Standard Deviation | 11.1 |
| Median | 49.5 | Range | 36 | | |

▶ **Guided**Practice

**2.** Find the mean, median, mode, range, and standard deviation of the data set obtained after multiplying each value by 0.8.

63, 47, 54, 60, 55, 46, 51, 60, 58, 50, 56, 60

**2 Comparing Distributions** Recall that when choosing appropriate statistics to represent data, you should first analyze the shape of the distribution. The same is true when comparing distributions.

- Use the mean and standard deviation to compare two symmetric distributions.

- Use the five-number summaries to compare two skewed distributions or a symmetric distribution and a skewed distribution.

### Example 3 Compare Data Using Histograms

**QUIZ SCORES Robert and Elaine's quiz scores for the first semester of Algebra 1 are shown below.**

| Robert's Quiz Scores |
|---|
| 85, 95, 70, 87, 78, 82, 84, 84, 85, 99, 88, 74, 75, 89, 79, 80, 92, 91, 96, 81 |

| Elaine's Quiz Scores |
|---|
| 89, 76, 87, 86, 92, 77, 78, 83, 83, 82, 81, 82, 84, 85, 85, 86, 89, 93, 77, 85 |

**a. Use a graphing calculator to construct a histogram for each set of data. Then describe the shape of each distribution.**

Enter Robert's quiz scores as **L1** and Elaine's quiz scores as **L2**.

**TechnologyTip**

Histograms To create a histogram for a set of data in **L2**, press [2nd] [STAT PLOT] [ENTER] [ENTER], choose ◧▥▥, and enter **L2** for **Xlist**.

Robert's Quiz Scores

[69, 101] scl: 4 by [0, 8] scl: 1

Elaine's Quiz Scores

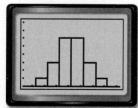

[69, 101] scl: 4 by [0, 8] scl: 1

Both distributions are high in the middle and low on the left and right. Therefore, both distributions are symmetric.

**b. Compare the data sets using either the means and standard deviations or the five-number summaries. Justify your choice.**

Both distributions are symmetric, so use the means and standard deviations to describe the centers and spreads.

Robert's Quiz Scores

Elaine's Quiz Scores

The means for the students' quiz scores are approximately equal, but Robert's quiz scores have a much higher standard deviation than Elaine's quiz scores. This means that Elaine's quiz scores are generally closer to her mean than Robert's quiz scores are to his mean.

▶ **Guided**Practice

**COMMUTE** The students in two of Mr. Martin's classes found the average number of minutes that they each spent traveling to school each day.

**3A.** Use a graphing calculator to construct a histogram for each set of data. Then describe the shape of each distribution.

**3B.** Compare the data sets using either the means and standard deviations or the five-number summaries. Justify your choice.

| 2nd Period (minutes) | 7th Period (minutes) |
|---|---|
| 8, 4, 18, 7, 13, 26, 12, 6, 20, 5, 9, 24, 8, 16, 31, 13, 17, 10, 8, 22, 12, 25, 13, 11, 18, 12, 16, 22, 25, 33 | 21, 4, 20, 13, 22, 6, 10, 23, 13, 25, 14, 16, 19, 21, 19, 8, 20, 18, 9, 14, 21, 17, 19, 22, 4, 19, 21, 26 |

Box-and-whisker plots are useful for comparisons of data because they can be displayed on the same screen.

● **Real-World Example 4** Compare Data Using Box-and-Whisker Plots

**FOOTBALL** Kurt's total rushing yards per game for his junior and senior seasons are shown.

| Junior Season (yards) | | | | | | Senior Season (yards) | | | | | |
|---|---|---|---|---|---|---|---|---|---|---|---|
| 16 | 20 | 72 | 4 | 25 | 18 | 77 | 54 | 109 | 60 | 156 | 72 |
| 34 | 10 | 42 | 17 | 56 | 12 | 39 | 83 | 73 | 101 | 46 | 80 |

**a. Use a graphing calculator to construct a box-and-whisker plot for each set of data. Then describe the shape of each distribution.**

Enter Kurt's rushing yards from his junior season as **L1** and his rushing yards from his senior season as **L2**. Graph both box-and-whisker plots on the same screen by graphing **L1** as **Plot1** and **L2** as **Plot2**.

For Kurt's junior season, the right whisker is longer than the left, and the median is closer to the left whisker. The distribution is positively skewed.

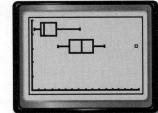

[0, 160] scl: 10 by [0, 5] scl: 1

For Kurt's senior season, the lengths of the whiskers are approximately equal, and the median is in the middle of the data. The distribution is symmetric.

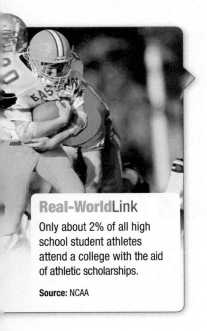

**b. Compare the data sets using either the means and standard deviations or the five-number summaries. Justify your choice.**

One distribution is symmetric and the other is skewed, so use the five-number summaries to compare the data.

The upper quartile for Kurt's junior season was 38, while the minimum for his senior season was 39. This means that Kurt rushed for more yards in every game during his senior season than 75% of the games during his junior season.

The maximum for Kurt's junior season was 72, while his median for his senior season was 75. This means that in half of his games during his senior year, he rushed for more yards than in any game during his junior season. Overall, we can conclude that Kurt rushed for many more yards during his senior season than during his junior season.

**StudyTip**

Box-and-Whisker Plots
Recall that a box-and-whisker plot displays the spread of a data set by dividing it into four quartiles. Each quartile accounts for 25% of the data.

▶ **Guided**Practice

**BASKETBALL**  The points Vanessa scored per game during her junior and senior seasons are shown.

**4A.** Use a graphing calculator to construct a histogram for each set of data. Then describe the shape of each distribution.

**4B.** Compare the data sets using either the means and standard deviations or the five-number summaries. Justify your choice.

| Junior Season (points) |
|---|
| 10, 12, 6, 10, 13, 8, 12, 3, 21, 14, 7, 0, 15, 6, 16, 8, 17, 3, 17, 2 |

| Senior Season (points) |
|---|
| 10, 32, 3, 22, 20, 30, 26, 24, 5, 22, 28, 32, 26, 21, 6, 20, 24, 18, 12, 25 |

## Check Your Understanding

**Example 1**  Find the mean, median, mode, range, and standard deviation of each data set that is obtained after adding the given constant to each value.

   **1.** 10, 13, 9, 8, 15, 8, 13, 12, 7, 8, 11, 12; + (−7)    **2.** 38, 36, 37, 42, 31, 44, 37, 45, 29, 42, 30, 42; + 23

**Example 2**  Find the mean, median, mode, range, and standard deviation of each data set that is obtained after multiplying each value by the given constant.

   **3.** 6, 10, 3, 7, 4, 9, 3, 8, 5, 11, 2, 1; × 3       **4.** 42, 39, 45, 44, 37, 42, 38, 37, 41, 49, 42, 36; × 0.5

**Example 3**  **5. TRACK**  Mark and Kyle's long jump distances are shown.

| Kyle's Distances (ft) |
|---|
| 17.2, 18.28, 18.56, 17.28, 17.36, 18.08, 17.43, 17.71, 17.46, 18.26, 17.51, 17.58, 17.41, 18.21, 17.34, 17.63, 17.55, 17.26, 17.18, 17.78, 17.51, 17.83, 17.92, 18.04, 17.91 |

| Mark's Distances (ft) |
|---|
| 18.88, 19.24, 17.63, 18.69, 17.74, 19.18, 17.92, 18.96, 18.19, 18.21, 18.46, 17.47, 18.49, 17.86, 18.93, 18.73, 18.34, 18.67, 18.56, 18.79, 18.47, 18.84, 18.87, 17.94, 18.7 |

**a.** Use a graphing calculator to construct a histogram for each set of data. Then describe the shape of each distribution.

**b.** Compare the data sets using either the means and standard deviations or the five-number summaries. Justify your choice.

**Example 4**

6. **TIPS** Miguel and Stephanie are servers at a restaurant. The tips that they earned to the nearest dollar over the past 15 workdays are shown.

| Miguel's Tips ($) |
|---|
| 14, 68, 52, 21, 63, 32, 43, 35, 70, 37, 42, 16, 47, 38, 48 |

| Stephanie's Tips ($) |
|---|
| 34, 52, 43, 39, 41, 50, 46, 36, 37, 47, 39, 49, 44, 36, 50 |

   a. Use a graphing calculator to construct a box-and-whisker plot for each set of data. Then describe the shape of each distribution.

   b. Compare the data sets using either the means and standard deviations or the five-number summaries. Justify your choice.

## Practice and Problem Solving

Extra Practice is on page R12.

**Example 1** Find the mean, median, mode, range, and standard deviation of each data set that is obtained after adding the given constant to each value.

   7. 52, 53, 49, 61, 57, 52, 48, 60, 50, 47; + 8    8. 101, 99, 97, 88, 92, 100, 97, 89, 94, 90; + (−13)

   9. 27, 21, 34, 42, 20, 19, 18, 26, 25, 33; + (−4)    10. 72, 56, 71, 63, 68, 59, 77, 74, 76, 66; + 16

**Example 2** Find the mean, median, mode, range, and standard deviation of each data set that is obtained after multiplying each value by the given constant.

   11. 11, 7, 3, 13, 16, 8, 3, 11, 17, 3; × 4    12. 64, 42, 58, 40, 61, 67, 58, 52, 51, 49; × 0.2

   13. 33, 37, 38, 29, 35, 37, 27, 40, 28, 31; × 0.8    14. 1, 5, 4, 2, 1, 3, 6, 2, 5, 1; × 6.5

**Example 3** 15. **BOOKS** The page counts for the books that the students chose are shown.

| 1st Period |
|---|
| 388, 439, 206, 438, 413, 253, 311, 427, 258, 511, 283, 578, 291, 358, 297, 303, 325, 506, 331, 482, 343, 372, 456, 267, 484, 227 |

| 6th Period |
|---|
| 357, 294, 506, 392, 296, 467, 308, 319, 485, 333, 352, 405, 359, 451, 378, 490, 379, 401, 409, 421, 341, 438, 297, 440, 500, 312, 502 |

   a. Use a graphing calculator to construct a histogram for each set of data. Then describe the shape of each distribution.

   b. Compare the data sets using either the means and standard deviations or the five-number summaries. Justify your choice.

16. **TELEVISIONS** The prices for a sample of televisions are shown.

| The Electronics Superstore |
|---|
| 46, 25, 62, 45, 30, 43, 40, 46, 33, 53, 35, 38, 39, 40, 52, 42, 44, 48, 50, 35, 32, 55, 28, 58 |

| Game Central |
|---|
| 53, 49, 26, 61, 40, 50, 42, 35, 45, 48, 31, 48, 33, 50, 35, 55, 38, 50, 42, 53, 44, 54, 48, 58 |

   a. Use a graphing calculator to construct a histogram for each set of data. Then describe the shape of each distribution.

   b. Compare the data sets using either the means and standard deviations or the five-number summaries. Justify your choice.

**Example 4** 17. **BRAINTEASERS** The time that it took Leon and Cassie to complete puzzles is shown.

| Leon's Times (minutes) |
|---|
| 4.5, 1.8, 3.2, 5.1, 2.0, 2.6, 4.8, 2.4, 2.2, 2.8, 1.8, 2.2, 3.9, 2.3, 3.3, 2.4 |

| Cassie's Times (minutes) |
|---|
| 2.3, 5.8, 4.8, 3.3, 5.2, 4.6, 3.6, 5.7, 3.8, 4.2, 5.0, 4.3, 5.5, 4.9, 2.4, 5.2 |

   a. Use a graphing calculator to construct a box-and-whisker plot for each set of data. Then describe the shape of each distribution.

   b. Compare the data sets using either the means and standard deviations or the five-number summaries. Justify your choice.

**18. DANCE** The total amount of money that a sample of students spent to attend the homecoming dance is shown.

| Boys (dollars) |
| --- |
| 114, 98, 131, 83, 91, 64, 94, 77, 96, 105, 72, 108, 87, 112, 58, 126 |

| Girls (dollars) |
| --- |
| 124, 74, 105, 133, 85, 162, 90, 109, 94, 102, 98, 171, 138, 89, 154, 76 |

   **a.** Use a graphing calculator to construct a box-and-whisker plot for each set of data. Then describe the shape of each distribution.

   **b.** Compare the data sets using either the means and standard deviations or the five-number summaries. Justify your choice.

**19. LANDSCAPING** Refer to the beginning of the lesson. Rhonda, another employee that works with Tom, earned the following over the past month.

| Rhonda's Pay ($) | | |
| --- | --- | --- |
| 45 | 55 | 53 |
| 47 | 53 | 54 |
| 44 | 56 | 59 |
| 63 | 47 | 53 |
| 60 | 57 | 62 |
| 44 | 50 | 45 |
| 60 | 53 | 49 |
| 62 | 47 | |

   **a.** Find the mean, median, mode, range, and standard deviation of Rhonda's earnings.

   **b.** A $5 bonus had been added to each of Rhonda's daily earnings. Find the mean, median, mode, range, and standard deviation of Rhonda's earnings before the $5 bonus.

**20. SHOPPING** The items Lorenzo purchased are shown.

   **a.** Find the mean, median, mode, range, and standard deviation of the prices.

   **b.** A 7% sales tax was added to the price of each item. Find the mean, median, mode, range, and standard deviation of the items without the sales tax.

| | |
| --- | --- |
| Baseball hat | $14.98 |
| Jeans | $24.61 |
| T-shirt | $12.84 |
| T-shirt | $16.05 |
| Backpack | $42.80 |
| Folders | $2.14 |
| Sweatshirt | $19.26 |

---

## H.O.T. Problems    Use Higher-Order Thinking Skills

**21. CHALLENGE** A salesperson has 15 SUVs priced between $33,000 and $37,000 and 5 luxury cars priced between $44,000 and $48,000. The average price for all of the vehicles is $39,250. The salesperson decides to reduce the prices of the SUVs by $2000 per vehicle. What is the new average price for all of the vehicles?

**22. REASONING** If every value in a set of data is multiplied by a constant $k$, $k < 0$, then how can the mean, median, mode, range, and standard deviation of the new data set be found?

**23. WRITING IN MATH** Compare and contrast the benefits of displaying data using histograms and box-and-whisker plots.

**24. REASONING** If $k$ is added to every value in a set of data, and then each resulting value is multiplied by a constant $m$, $m > 0$, how can the mean, median, mode, range, and standard deviation of the new data set be found? Explain your reasoning.

**25. WRITING IN MATH** Explain why the mean and standard deviation are used to compare the center and spread of two symmetrical distributions and the five-number summary is used to compare the center and spread of two skewed distributions or a symmetric distribution and a skewed distribution.

# 23 Graphing Technology Lab
# The Normal Curve

When there are a large number of values in a data set, the frequency distribution tends to cluster around the mean of the set in a distribution (or shape) called a **normal distribution**. The graph of a normal distribution is called a **normal curve**. Since the shape of the graph resembles a bell, the graph is also called a *bell curve*.

Data sets that have a normal distribution include reaction times of drivers that are the same age, achievement test scores, and the heights of people that are the same age.

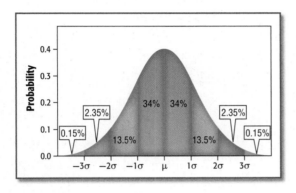

You can use a graphing calculator to graph and analyze a normal distribution if the mean and standard deviation of the data are known.

## Activity 1 Graph a Normal Distribution

**HEIGHT** The mean height of 15-year-old boys in the city where Isaac lives is 67 inches, with a standard deviation of 2.8 inches. Use a normal distribution to represent these data.

**Step 1** Set the viewing window. [WINDOW]

- Xmin = 67 [−] 3 [×] 2.8 [ENTER] **58.6**
- Xmax = 67 [+] 3 [×] 2.8 [ENTER] **75.4**
- Xscl = 2.8 [ENTER]
- Ymin = 0 [ENTER]
- Ymax = 1 [÷] [(] 2 [×] 2.8 [)] [ENTER] **.17857142...**
- Yscale = 1 [ENTER]

**Step 2** By entering the mean and standard deviation into the calculator, we can graph the corresponding normal curve. Enter the values using the following keystrokes.

**KEYSTROKES:** [Y=] [2nd] [DISTR] [ENTER] [X,T,θ,n] [,] 67 [,] 2.8 [)] [GRAPH]

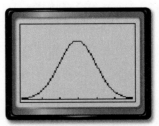

[58.6, 75.4] scl: 2.8 by [0, 0.17857142] scl: 1

*(continued on the next page)*

The probability of a range of values is the area under the curve.

## Activity 2 Analyze a Normal Distribution

**Use the graph to answer questions about the data. What is the probability that Isaac will be at most 67 inches tall when he is 15?**

The sum of all the $y$-values up to $x = 67$ would give us the probability that Isaac's height will be less than or equal to 67 inches. This is also the area under the curve. We will shade the area under the curve from negative infinity to 67 inches and find the area of the shaded portion of the graph.

**Step 1** Use the **ShadeNorm** function.

KEYSTROKES: [2nd] [DISTR] [▶] [ENTER]

**Step 2** Shade the graph.

Next enter the lowest value, highest value, mean, and standard deviation.

On the TI-84 Plus, $-1 \times 10^{99}$ represents negative infinity.

KEYSTROKES: [(−)] 1 [2nd] [EE] 99 [,] 67 [,] 67 [,] 2.8 [)] [ENTER]

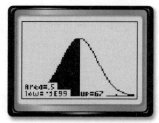

[58.6, 75.4] scl: 2.8 by [0, 0.17857142] scl: 1

The area is given as 0.5. The probability that Isaac will be 67 inches tall is 0.5 or 50%. Since the mean value is 67, we expect the probability to be 50%.

## Exercises

1. What is the probability that Isaac will be at least 6 feet tall when he is 15?

2. What is the probability that Isaac will be between 65 and 68 inches?

3. The **z-score** represents the number of standard deviations that a given data value is from the mean. The z-score for a data value $X$ is given by $z = \dfrac{X - \mu}{\sigma}$, where $\mu$ is the mean and $\sigma$ is the standard deviation. Find and interpret the z-score of a height of 73 inches.

4. Find and interpret the z-score of a height of 61 inches.

## Extension

**Refer to the curve at the right.**

5. Compare this curve to the normal curve in Activity 1.

6. Describe where an outlier of the data set would be graphed on this curve.

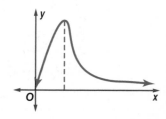

# 24 Algebra Lab
# Two-Way Frequency Tables

Joana sent out a survey to the freshmen and sophomores, asking if they were planning on attending the dance. One way of organizing her responses is to use a two-way frequency table. A **two-way frequency table** or *contingency table* is used to show the frequencies of data from a survey or experiment classified according to two categories, with the rows indicating one category and the columns indicating the other.

For Joana's survey, the two categories are *class* and *attendance*. These categories can be split into subcategories: *freshman* and *sophomore* for *class*, and *attending* and *not attending* for *attendance*.

| Class | Attending | Not Attending | Totals |
|---|---|---|---|
| Freshman | | | |
| Sophomore | | | |
| Totals | | | |

subcategories

## Activity 1 Two-Way Frequency Table

**DANCE** Sixty-six freshmen responded to the survey, with 32 saying that they would be attending. Of the 84 sophomores that responded, 46 said they would attend. Organize the data in a two-way table.

**Step 1** Find the values for every combination of subcategories. One combination is freshmen/not attending. Since 32 of 66 freshmen are attending, 66 − 32 or 34 freshmen are *not* attending. These combinations are called **joint frequencies.**

**Step 2** Place every combination in the corresponding cell.

**Step 3** Find the totals of each subcategory and place them in their corresponding cell. These values are called **marginal frequencies.**

**Step 4** Find the sum of each set of marginal frequencies. These two sums should be equal. Place the value in the bottom right corner.

| Class | Attending | Not Attending | Totals |
|---|---|---|---|
| Freshman | 32 | 34 | 66 |
| Sophomore | 46 | 38 | 84 |
| Totals | 78 | 72 | 150 |

joint frequencies

marginal frequencies

marginal frequencies

## Analyze the Results

1. How many students responded to the survey?

2. How many of the students that were surveyed are attending the dance?

3. How many of the surveyed sophomores are not attending the dance?

4. What does each of the joint frequencies represent?

5. What does each of the marginal frequencies represent?

6. **WORK** Heather sent out a survey asking who was working during the holiday. Of the 50 boys who responded, 34 said *yes*. Of the 45 girls who responded, 21 said *no*. Create a two-way frequency table of the results.

7. **SOCCER** Pamela asked if anyone would be interested in a co-ed soccer team. Of the 28 boys who responded, 18 said that they would play and 4 were undecided. Of the 22 girls who responded, 6 said they did not want to play and 3 were undecided. Create a two-way frequency table of the results.

A **relative frequency** is the ratio of the number of observations in a category to the total number of observations. Relative frequencies are also probabilities. To create a relative frequency two-way table, divide each of the values by the total number of observations and replace them with their corresponding decimals or percents.

| Class | Attending | Not Attending | Totals |
|---|---|---|---|
| Freshman | $\frac{32}{150} \approx 21.3\%$ | 22.7% | 44% |
| Sophomore | 30.7% | 25.3% | 56% |
| **Totals** | 52% | 48% | 100% |

A **conditional relative frequency** is the ratio of the joint frequency to the marginal frequency. For example, given that a student is a freshman, what is the conditional relative frequency that he or she is going to the dance? In other words, what is the probability that a freshman is going to the dance?

### Activity 2 Two-Way Conditional Relative Frequency Table

**DANCE Joana wants to determine the conditional relative frequencies (or probabilities) given the fact that she knows the class of the respondents.**

**Step 1** Refer to the table in Activity 1. A total of 66 freshmen responded, and 32 said *yes*. Therefore, the conditional relative frequency that a respondent said *yes* given that the respondent is a freshman is $\frac{32}{66}$.

**Step 2** Place every conditional relative frequency in the corresponding cell.

**Step 3** The conditional relative frequencies for each row should sum to 100%.

| Conditional Relative Frequencies by Class | | | |
|---|---|---|---|
| Class | Attending | Not Attending | Totals |
| Freshman | $\frac{32}{66} \approx 48\%$ | $\frac{34}{66} \approx 52\%$ | 100% |
| Sophomore | $\frac{46}{84} \approx 55\%$ | $\frac{38}{84} \approx 45\%$ | 100% |

## Analyze the Results

8. Given that a respondent was a sophomore, what is the probability that he or she said *no*?

9. What does each of the conditional relative frequencies represent?

10. Why do you think that the columns do not sum to 100%?

11. Create a two-way conditional relative frequency table for the category *attendance*.

12. Given that a respondent was not attending, what is the probability that he or she is a freshman?

13. **ACTIVITIES** The managers, staff, and assistants were given three options for the holiday activity: a potluck, a dinner at a restaurant, and a gift exchange. Five of the 11 managers want a dinner, while 3 want a potluck. Eleven of the 45 staff members want a gift exchange, while 18 want a dinner. Ten of the 32 assistants want a dinner, while 8 of them want a gift exchange.

   a. Create a two-way frequency table.

   b. Convert the two-way frequency table into a relative frequency table.

   c. Create two conditional relative frequency tables: one for the activities and one for the employees.

## Additional Exercises

### Use with Lesson 1-1.

1. **SMARTPHONES** A certain smartphone family plan costs $55 per month plus additional usage costs. If $x$ is the number of cell phone minutes used above the plan amount and $y$ is the number of megabytes of data used above the plan amount, interpret the following expressions.
   a. $0.25x$
   b. $2y$
   c. $0.25x + 2y + 55$

### Use with Lesson 1-2.

2. **SPORTS** Kamilah sells tickets at Duke University's athletic ticket office. If $p$ represents a preferred season ticket, $b$ represents a blue zone ticket, and $g$ represents a general admission ticket, interpret and then evaluate the following expressions.
   a. $45b$
   b. $15p + 35g$
   c. $6p + 11b + 22g$

### Use with Lesson 1-3.

3. **RETAIL** The table shows prices on children's clothing.

   | Shorts | Shirts | Tank Tops |
   |--------|--------|-----------|
   | $7.99  | $8.99  | $6.99     |
   | $5.99  | $4.99  | $2.99     |

   a. Interpret the expression $5(8.99) + 2(2.99) + 7(5.99)$.
   b. Write and evaluate three different expressions that represent 8 pairs of shorts and 8 tops.
   c. If you buy 8 shorts and 8 tops, you receive a discount of 15%. Find the greatest and least amount of money you can spend on the 16 items at the sale.

### Use with Lesson 1-4.

4. **FOOD** Kenji is picking up take-out food for his study group.

   | Menu | |
   |------|--------|
   | Item | Cost ($) |
   | sandwich | 2.49 |
   | cup of soup | 1.29 |
   | side salad | 0.99 |
   | drink | 1.49 |

   a. Interpret the expression $4(2.49) + 3(1.29) + 3(0.99) + 5(1.49)$.
   b. How much would it cost if Kenji bought four of each item on the menu?

### Use with Lesson 1-7.

5. **CELL PHONE PICTURES** The cost of sending cell phone pictures is given by $y = 0.25x$, where $x$ is the number of pictures that you send.
   a. Write the equation in function notation. Interpret the function in terms of the context.
   b. Find $f(5)$ and $f(12)$. What do these values represent?
   c. Determine the domain and range of this function.

6. **EDUCATION** The average national math test scores $f(t)$ for 17-year-olds can be represented as a function of the national science scores $t$ by $f(t) = 0.8t + 72$.
   a. Graph this function. Interpret the function in terms of the context.
   b. What is the science score that corresponds to a math score of 308?
   c. What is the domain and range of this function?

7. **ERROR ANALYSIS** Corazon thinks $f(x)$ and $g(x)$ are representations of the same function. Maggie disagrees. Who is correct? Explain your reasoning.

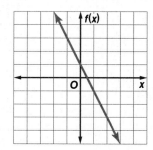

   | x | g(x) |
   |----|------|
   | −1 | 1 |
   | 0 | −1 |
   | 1 | −3 |
   | 2 | −5 |
   | 3 | −7 |

### Use with Lesson 2-2.

8. **CHALLENGE** Solve each equation for $x$. Assume that $a \neq 0$.
   a. $ax = 12$
   b. $x + a = 15$
   c. $-5 = x - a$
   d. $\frac{1}{a}x = 10$

### Use with Explore 2-3.

9. **CHALLENGE** Solve each equation for $x$. Assume that $a \neq 0$.
   a. $ax + 7 = 5$
   b. $\frac{1}{a}x - 4 = 9$
   c. $2 - ax = -8$

### Use with Lesson 2-4.

10. **REASONING** Solve $5x + 2 = ax - 1$ for $x$. Assume that $a \neq 0$. Describe each step.

## Additional Exercises

### Use with Lesson 3-5.

**11. SPORTS** To train for an upcoming marathon, Olivia plans to run 3 miles per day for the first week, and then increase the daily distance by a half a mile each of the following weeks.

  **a.** Write an equation to represent the $n$th term of the sequence.

  **b.** If the pattern continues, during which week will she run 10 miles per day?

  **c.** Is it reasonable to think that this pattern will continue indefinitely? Explain.

### Use with Lesson 3-6.

**12. ERROR ANALYSIS** Quentin thinks that $f(x)$ and $g(x)$ are both proportional. Claudia thinks they are not proportional. Is either of them correct? Explain your reasoning.

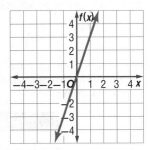

| x | g(x) |
|----|------|
| −2 | −7 |
| −1 | −4 |
| 0 | −1 |
| 1 | 2 |
| 2 | 5 |

### Use with Explore 4-2.

**13. COMBINING FUNCTIONS** The parents of a college student open an account for her with a deposit of $5000, and they set up automatic deposits of $100 to the account every week.

  **a.** Write a function $d(t)$ to express the amount of money in the account $t$ weeks after the initial deposit.

  **b.** The student plans on spending $600 the first week and $250 in each of the following weeks for room and board and other expenses. Write a function $w(t)$ to express the amount of money taken out of the account each week.

  **c.** Find $B(t) = d(t) - w(t)$. What does this new function represent?

  **d.** Will the student run out of money? If so, when?

### Use with Lesson 4-3.

Write an equation for the line described in standard form.

**14.** through $(-1, 7)$ and $(8, -2)$

**15.** through $(-4, 3)$ with $y$-intercept 0

**16.** with $x$-intercept 4 and $y$-intercept 5

**17. ERROR ANALYSIS** Juana thinks that $f(x)$ and $g(x)$ have the same slope but different intercepts. Sabrina thinks that $f(x)$ and $g(x)$ describe the same line. Is either of them correct? Explain your reasoning.

> The graph of $g(x)$ is the line that passes through $(3, -7)$ and $(-6, 4)$.

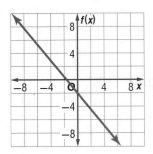

### Use with Lesson 4-4.

**18. REASONING** Which key features of the graphs of two parallel lines are the same, and which are different? Which key features of the graphs of two perpendicular lines are the same, and which are different?

### Use with Explore 5-2.

**19. CHALLENGE** Solve each inequality for $x$. Assume that $a > 0$.

  **a.** $-ax < 5$

  **b.** $\frac{1}{a}x \geq 8$

  **c.** $-6 \geq ax$

### Use with Lesson 5-3.

**20. CHALLENGE** Solve each inequality for $x$. Assume that $a > 0$.

  **a.** $ax + 4 \geq -ax - 5$

  **b.** $2 - ax < x$

  **c.** $-\frac{2}{a}x + 3 > -9$

### Use with Explore 5-4.

**21. CHALLENGE** Solve each inequality for $x$. Assume $a$ is constant and $a > 0$.

  **a.** $-3 < ax + 1 \leq 5$

  **b.** $-\frac{1}{a}x + 6 < 1$ or $2 - ax > 8$

## Use with Lesson 5-6.

22. **CHALLENGE** Write a linear inequality for which $(-1, 2)$, $(1, 1)$, and $(3, -4)$ are solutions but $(0, 1)$ is not.

## Use with Lesson 7-2.

23. **PROBABILITY** The probability of rolling a die and getting an even number is $\frac{1}{2}$. If you roll the die twice, the probability of getting an even number both times is $\left(\frac{1}{2}\right)\left(\frac{1}{2}\right)$ or $\left(\frac{1}{2}\right)^2$.

   a. What does $\left(\frac{1}{2}\right)^4$ represent?

   b. Write an expression to represent the probability of rolling a die $d$ times and getting an even number every time. Write the expression as a power of 2.

## Use with Lesson 7-3.

24. **SMARTPHONES** A recent cell phone study showed that company A's phone processes up to $7.95 \times 10^5$ bits of data every second. Company B's phone processes up to $1.41 \times 10^6$ bits of data every second. Evaluate and interpret $\frac{1.41 \times 10^6}{7.95 \times 10^5}$.

25. **HEALTH** A ponderal index $p$ is a measure of a person's body based on height $h$ in meters and mass $m$ in kilograms. One such formula is $p = 100m^{\frac{1}{3}}h^{-1}$. If a person who is 182 centimeters tall has a ponderal index of about 2.2, how much does the person weigh in kilograms?

## Use with Lesson 9-1.

26. **ERROR ANALYSIS** Jade thinks that the parabolas represented by the graph and the description have the same axis of symmetry. Chase disagrees. Who is correct? Explain your reasoning.

> a parabola that opens downward, passing through (0, 6) and having a vertex at (2, 2)

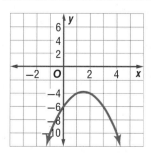

27. **WRITING IN MATH** Use tables and graphs to compare and contrast an exponential function $f(x) = ab^x + c$, where $a \neq 0$, $b > 0$, and $b \neq 1$, a quadratic function $g(x) = ax^2 + c$, and a linear function $h(x) = ax + c$. Include intercepts, portions of the graph where the functions are increasing, decreasing, positive, or negative, relative maxima and minima, symmetries, and end behavior. Which function eventually exceeds the others?

## Use with Lesson 9-5.

28. **MULTIPLE REPRESENTATIONS** In this problem, you will investigate writing a quadratic equation with given roots. If $p$ is a root of $0 = ax^2 + bx + c$, then $(x - p)$ is a factor of $ax^2 + bx + c$.

   a. **TABULAR** Copy and complete the first two columns of the table.

   | Roots | Factors | Equation |
   |---|---|---|
   | $2, 5$ | $(x - 2), (x - 5)$ | $(x - 2)(x - 5) = 0$ <br> $x^2 - 7x + 10 = 0$ |
   | $1, 9$ | | |
   | $-1, 3$ | | |
   | $0, 6$ | | |
   | $\frac{1}{2}, 7$ | | |
   | $-\frac{2}{3}, 4$ | | |

   b. **ALGEBRAIC** Multiply the factors to write each equation with integral coefficients. Use the equations to complete the last column of the table. Write each equation.

   c. **ANALYTICAL** How could you write an equation with three roots? Test your conjecture by writing an equation with roots 1, 2, and 3. Is the equation quadratic? Explain.

## Use with Lesson 9-6.

29. **FINANCIAL LITERACY** Daniel deposited $500 into a savings account and after 8 years, his investment is worth $807.07. The equation $A = d(1.005)^{12t}$ models the value of Daniel's investment $A$ after $t$ years with an initial deposit $d$.

   a. What would the value of Daniel's investment be if he had deposited $1000?

   b. What would the value of Daniel's investment be if he had deposited $250?

   c. Interpret $d(1.005)^{12t}$ to explain how the amount of the original deposit affects the value of Daniel's investment.

30. **REASONING** Use tables and graphs to compare and contrast an exponential function $f(x) = ab^x + c$, where $a \neq 0$, $b > 0$, and $b \neq 1$, and a linear function $g(x) = ax + c$. Include intercepts, intervals where the functions are increasing, decreasing, positive, or negative, relative maxima and minima, symmetry, and end behavior.

**31. WRITING IN MATH** Compare and contrast the graphs of absolute value, step, and piecewise-defined functions with the graphs of quadratic and exponential functions. Discuss the domains, ranges, maxima, minima, and symmetry.

## Use with Lesson 9-7.

**32. COMBINING FUNCTIONS** A swimming pool is losing water at a rate of 0.5% per hour. The maximum amount of water in the pool is 20,500 gallons.

  **a.** Write an exponential function $w(t)$ to express the amount of water in the pool after time $t$. Assume that the pool is at maximum capacity at $t = 0$.

  **b.** A pump sends water into a pool whenever the level of water in the pool drops below 19,000 gallons. It then pumps 1500 gallons of water into the pool over 30 minutes. Write a function $p(t)$ where $t$ is time in hours to express the rate at which the water is pumped into the pool.

  **c.** Use the graph of $p(t)$ to determine when the pump turn on the first time.

  **d.** Find $C(t) = p(t) + w(t)$. What does this new function represent?

## Use with Lesson 9-9.

**33. PROOF** Write a paragraph proof to show that linear functions grow by equal differences over equal intervals, and exponential functions grow by equal factors over equal intervals. (*Hint:* Let $y = ax$ represent a linear function and let $y = a^x$ represent an exponential function.)

## Use with Lesson 10-3.

**34. REASONING** Make a conjecture about the sum of a rational number and an irrational number. Is the sum *rational* or *irrational*? Is the product of a nonzero rational number and an irrational number *rational* or *irrational*? Explain your reasoning.

# Lesson 2 / Practice

## Interpreting Graphs of Functions

Identify the function graphed as *linear* or *nonlinear*. Then estimate and interpret the intercepts of the graph, any symmetry, where the function is positive, negative, increasing, and decreasing, the *x*-coordinate of any relative extrema, and the end behavior of the graph.

**1.**

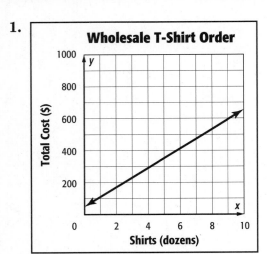

**2.**

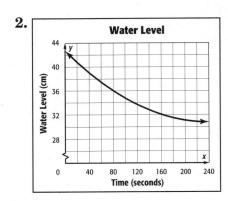

**3.**

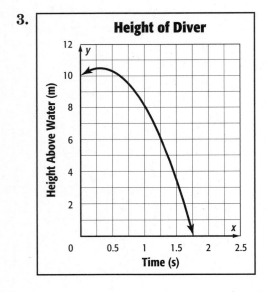

**4.**

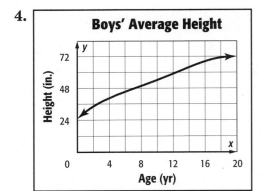

## Lesson 5 Practice

### Regression and Median-Fit Lines

**Write an equation of the regression line for the data in each table below. Then find the correlation coefficient.**

**1. TURTLES** The table shows the number of turtles hatched at a zoo each year since 2006.

| Year | 2006 | 2007 | 2008 | 2009 | 2010 |
|---|---|---|---|---|---|
| Turtles Hatched | 21 | 17 | 16 | 16 | 14 |

**2. SCHOOL LUNCHES** The table shows the percentage of students receiving free or reduced price school lunches at a certain school each year since 2006.

| Year | 2006 | 2007 | 2008 | 2009 | 2010 |
|---|---|---|---|---|---|
| Percentage | 14.4% | 15.8% | 18.3% | 18.6% | 20.9% |

**Source:** KidsData

**3. SPORTS** Below is a table showing the number of students signed up to play lacrosse after school in each age group.

| Age | 13 | 14 | 15 | 16 | 17 |
|---|---|---|---|---|---|
| Lacrosse Players | 17 | 14 | 6 | 9 | 12 |

**4. LANGUAGE** The State of California keeps track of how many millions of students are learning English as a second language each year.

| Year | 2003 | 2004 | 2005 | 2006 | 2007 |
|---|---|---|---|---|---|
| English Learners | 1.600 | 1.599 | 1.592 | 1.570 | 1.569 |

**Source:** California Department of Education

**a.** Find an equation for the median-fit line.

**b.** Predict the number of students who were learning English in California in 2001.

**c.** Predict the number of students who were learning English in California in 2010.

**5. POPULATION** Detroit, Michigan, like a number of large cities, is losing population every year. Below is a table showing the population of Detroit each decade.

| Year | 1960 | 1970 | 1980 | 1990 | 2000 |
|---|---|---|---|---|---|
| Population (millions) | 1.67 | 1.51 | 1.20 | 1.03 | 0.95 |

**Source:** U.S. Census Bureau

**a.** Find an equation for the regression line.

**b.** Find the correlation coefficient and explain the meaning of its sign.

**c.** Estimate the population of Detroit in 2008.

## Lesson 6 Practice

### Inverse Linear Functions

**Find the inverse of each relation.**

**1.** {(−2, 1), (−5, 0), (−8, −1), (−11, 2)}

**2.** {(3, 5), (4, 8), (5, 11), (6, 14)}

**3.** {(5, 11), (1, 6), (−3, 1), (−7, −4)}

**4.** {(0, 3), (2, 3), (4, 3), (6, 3)}

**Graph the inverse of each function.**

**5.**
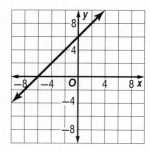

**6.**
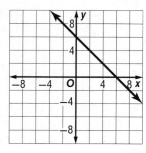

**7.**
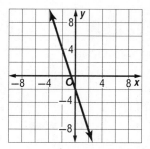

**Find the inverse of each function.**

**8.** $f(x) = \dfrac{6}{5}x - 3$

**9.** $f(x) = \dfrac{4x + 2}{3}$

**10.** $f(x) = \dfrac{3x - 1}{6}$

**11.** $f(x) = 3(3x + 4)$

**12.** $f(x) = -5(-x - 6)$

**13.** $f(x) = \dfrac{2x - 3}{7}$

**Write the inverse of each equation in $f^{-1}(x)$ notation.**

**13.** $4x + 6y = 24$

**14.** $-3y + 5x = 18$

**15.** $x + 5y = 12$

**16.** $5x + 8y = 40$

**17.** $-4y - 3x = 15 + 2y$

**18.** $2x - 3 = 4x + 5y$

**19. CHARITY** Jenny is running in a charity event. One donor is paying an initial amount of $20.00 plus an extra $5.00 for every mile that Jenny runs.

   **a.** Write a function $D(x)$ for the total donation for $x$ miles run.

   **b.** Find the inverse function, $D^{-1}(x)$.

   **c.** What do $x$ and $D^{-1}(x)$ represent in the context of the inverse function?

## Lesson 8 / Practice

### *Rational Exponents*

**Write each expression in radical form, or write each radical in exponential form.**

**1.** $\sqrt{13}$

**2.** $\sqrt{37}$

**3.** $\sqrt{17x}$

**4.** $(7ab)^{\frac{1}{2}}$

**5.** $21z^{\frac{1}{2}}$

**6.** $13(ab)^{\frac{1}{2}}$

**Simplify.**

**7.** $\left(\dfrac{1}{81}\right)^{\frac{1}{4}}$

**8.** $\sqrt[5]{1024}$

**9.** $512^{\frac{1}{3}}$

**10.** $\left(\dfrac{32}{1024}\right)^{\frac{1}{5}}$

**11.** $\sqrt[4]{1296}$

**12.** $3125^{\frac{1}{5}}$

**Solve each equation.**

**13.** $3^x = 729$

**14.** $4^x = 4096$

**15.** $5^x = 15{,}625$

**16.** $6^{x+3} = 7776$

**17.** $3^{x-3} = 2187$

**18.** $4^{3x+4} = 16{,}384$

**19. WATER** The flow of water $F$ in cubic feet per second over a wier, a small overflow dam, can be represented by $F = 1.26H^{\frac{3}{2}}$, where $H$ is the height of the water in meters above the crest of the wier. Find the height of the water if the flow of the water is 10.08 cubic feet per second.

# Lesson 10  Practice

## Transformations of Quadratic Functions

**Describe how the graph of each function is related to the graph of $f(x) = x^2$.**

**1.** $g(x) = (10 + x)^2$

**2.** $g(x) = -\dfrac{2}{5} + x^2$

**3.** $g(x) = 9 - x^2$

**4.** $g(x) = 2x^2 + 2$

**5.** $g(x) = -\dfrac{3}{4}x^2 - \dfrac{1}{2}$

**6.** $g(x) = -3(x + 4)^2$

**Match each equation to its graph.**

**A.** 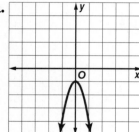   **B.** 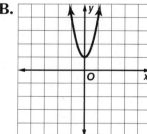   **C.**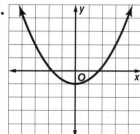

**7.** $y = -3x^2 - 1$

**8.** $y = \dfrac{1}{3}x^2 - 1$

**9.** $y = 3x^2 + 1$

**List the functions in order from the most vertically stretched to the least vertically stretched graph.**

**10.** $f(x) = 3x^2$, $g(x) = \dfrac{1}{2}x^2$, $h(x) = -2x^2$

**11.** $f(x) = \dfrac{1}{2}x^2$, $g(x) = -\dfrac{1}{6}x^2$, $h(x) = 4x^2$

**12. PARACHUTING** Two parachutists jump at the same time from two different planes as part of an aerial show. The height $h_1$ of the first parachutist in feet after $t$ seconds is modeled by the function $h_1 = -16t^2 + 5000$. The height $h_2$ of the second parachutist in feet after $t$ seconds is modeled by the function $h_2 = -16t^2 + 4000$.

**a.** What is the parent function of the two functions given?

**b.** Describe the transformations needed to obtain the graph of $h_1$ from the parent function.

**c.** Which parachutist will reach the ground first?

# Lesson 16 Practice

## Recursive Formulas

**Find the first five terms of each sequence.**

**1.** $a_1 = 25$, $a_n = a_{n-1} - 12$, $n \geq 2$

**2.** $a_1 = -101$, $a_n = a_{n-1} + 38$, $n \geq 2$

**3.** $a_1 = 3.3$, $a_n = a_{n-1} + 2.7$, $n \geq 2$

**4.** $a_1 = 7$, $a_n = -3a_{n-1} + 20$, $n \geq 2$

**5.** $a_1 = 20$, $a_n = \frac{1}{5}a_{n-1}$, $n \geq 2$

**6.** $a_1 = \frac{2}{3}$, $a_n = \frac{1}{3}a_{n-1} - \frac{2}{9}$, $n \geq 2$

**Write a recursive formula for each sequence.**

**7.** 80, −40, 20, −10, …

**8.** 87, 52, 17, −18, …

**9.** $\frac{1}{3}, \frac{4}{15}, \frac{16}{75}, \frac{64}{375}, \ldots$

**10.** $\frac{4}{5}, \frac{3}{10}, -\frac{1}{5}, -\frac{7}{10}, \ldots$

**11.** 2.6, 5.2, 7.8, 10.4, …

**12.** 100, 120, 144, 172.8, …

**13. PIZZA** The total costs for ordering one to five cheese pizzas from Luigi's Pizza Palace are shown.

**a.** Write a recursive formula for the sequence.

**b.** Write an explicit formula for the sequence.

| Total Number of Pizzas Ordered | Cost |
|---|---|
| 1 | $7.00 |
| 2 | $12.50 |
| 3 | $18.00 |
| 4 | $23.50 |
| 5 | $29.00 |

# Lesson 21 / Practice

## Distributions of Data

Use a graphing calculator to construct a histogram for the data, and use it to describe the shape of the distribution. **1–3. See students' graphs.**

**1.** 52, 42, 46, 53, 22, 36, 49, 23, 50, 44, 25, 28, 48, 45, 54, 50, 18, 38

40, 34, 53, 42, 16, 44, 50, 42, 45, 50, 25, 47, 33, 48, 49, 36, 49, 39

**2.** 51, 44, 54, 48, 63, 57, 58, 46, 55, 51, 63, 52, 46, 56, 57, 48, 52, 49

50, 56, 61, 51, 45, 52, 53, 55, 62, 55, 50, 53, 60, 56, 57, 59, 54, 45

**3.** 42, 19, 24, 14, 55, 21, 51, 36, 22, 16, 32, 18, 46, 49, 64, 12, 19, 39

17, 20, 35, 52, 23, 17, 25, 33, 18, 26, 17, 24, 13, 27, 37, 29, 30, 19

Describe the center and spread of the data using either the mean and standard deviation or the five-number summary. Justify your choice by constructing a box-and-whisker plot for the data.

**4.** 78, 82, 76, 48, 71, 78, 65, 78, 81, 76, 53, 63, 79, 60, 78, 59, 78, 61 70, 68, 70, 58, 45, 72, 78, 86, 73, 77, 80, 60, 75, 84, 67, 79, 70, 75

**5.** 63, 46, 48, 41, 72, 54, 48, 57, 53, 80, 52, 64, 55, 44, 67, 45, 71, 48 61, 45, 74, 49, 69, 54, 50, 72, 66, 50, 44, 58, 60, 54, 48, 59, 43, 70

**6.** 33, 25, 18, 46, 35, 25, 18, 39, 33, 44, 20, 31, 39, 24, 24, 26, 15, 28 23, 29, 40, 19, 20, 31, 45, 37, 30, 17, 38, 21, 43, 14, 30, 47, 42, 34

**7. GASOLINE** The average prices per gallon of gasoline during the first week of August on the east coast for the past 18 years are shown. Describe the center and spread of the data using either the mean and standard deviation or the five-number summary. Justify your choice by creating a box-and-whisker plot for the data.

| Price per Gallon (dollars) |
| --- |
| 1.05, 1.09, 1.13, 1.17, 1.15, 0.99, 1.12, 1.44, 1.28, 1.34, 1.49, 1.85, 2.26, 3.00, 2.81, 3.87, 2.51, 2.66 |

## Lesson 22 Practice

### Comparing Sets of Data

**Find the mean, median, mode, range, and standard deviation of each data set that is obtained after adding the given constant to each value.**

**1.** 62, 58, 57, 65, 68, 71, 49, 48, 52, 47; + 5.8    **2.** 2, 8, 1, 5, 1, 3, 1, 7, 5, 4, 3, 1; + (−0.3)

**3.** 4.3, 3.8, 3.1, 4.5, 3.7, 4.4, 4.9, 3.9; + (−2.4)    **4.** 17, 21, 18, 32, 29, 24, 19, 32; + 7.6

**Find the mean, median, mode, range, and standard deviation of each data set that is obtained after multiplying each value by the given constant.**

**5.** 94, 90, 88, 92, 85, 92, 86, 98, 92, 90; × 0.8    **6.** 41, 44, 47, 40, 43, 41, 42, 48; × 2.3

**7.** 63, 62, 59, 68, 67, 72, 70, 75, 64, 61; × $\frac{1}{3}$    **8.** 9, 7, 5, 2, 8, 4, 5, 6, 9, 5, 2, 1; × $\frac{4}{9}$

**9. RECYCLING** The weekly totals of recycled paper in pounds for two neighboring high schools are shown below.

| Highland Heights High School |
|---|
| 86, 57, 52, 43, 48, 55, 47, 64, 51, 77, 50, 62, 74, 70, 68, 53, 81, 53 |

| Valley Forge High School |
|---|
| 68, 79, 58, 101, 83, 65, 47, 73, 62, 77, 49, 84, 103, 70, 54, 97, 88, 94 |

**a.** Use a graphing calculator to construct a box-and-whisker plot for each set of data. Then describe the shape of each distribution.

**b.** Compare the data sets using either the means and standard deviations or the five-number summaries. Justify your choice.

# Common Core Standards Practice and Review

Boston, Massachusetts    Chandler, Arizona    Glenview, Illinois    Upper Saddle River, New Jersey

ISBN-13:  978-0-13-318567-6
ISBN-10:   0-13-318567-2

1 2 3 4 5 6 7 8 9 10  V001  19 18 17 16 15 14 13

# Table of Contents

# Common Core Standards Practice                    Week 1

## Selected Response

**1.** How many terms are in the expression $4a + a^2 + 6a^3$?

   **A** 3

   **B** 4

   **C** 5

   **D** 7

**2.** A sporting complex charges $5 to use its facility. The expression $0.15b + 5$ models the total cost to hit $b$ baseballs in the batting cages. What is the cost per baseball?

   **A** $5.15

   **B** $5.00

   **C** $.15

   **D** $.10

## Constructed Response

**3.** What is the value of the expression $2(ab)^3 - 3a + 5b$ for $a = 3$ and $b = 5$?

## Extended Response

**4.** A student stated that the sum of two real numbers is always an integer. Is the student correct? Explain why or why not and provide an example to support your answer.

# Common Core Standards Practice

# Week 2

## Selected Response

1. Which equation represents the following sentence?

   The sum of a number $n$ and 9 is 17.

   **A** $n + 9 = 17$
   **B** $9n = 17$
   **C** $n - 9 = 17$
   **D** $n + 17 = 9$

2. Which ordered pair is a solution to the equation $y = 2x + 8$ ?

   **A** $(2, 1)$
   **B** $(-2, -3)$
   **C** $(0, 8)$
   **D** $(8, 0)$

## Constructed Response

3. Allie earns $5 per hour doing chores.

   **a.** Make a table and write an equation to show the relationship between the number of hours worked $h$ and the wages earned $w$.

   **b.** How many hours will Allie need to work to earn $25?

## Extended Response

4. Consider the ordered pairs (0, 55), (1, 60), (2, 65), (3, 70), (4, 75), and (5, 80).

   **a.** Represent the ordered pairs as a table.

   **b.** Represent the ordered pairs as an equation.

   **c.** Represent the ordered pairs as a graph.

   **d.** Describe a situation that ordered pairs might represent.

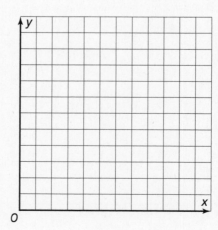

# Common Core Standards Practice          **Week 3**

**Selected Response**

1. What is the solution of $-36 = \frac{q}{4}$?

   A  $-144$

   B  $-9$

   C  $9$

   D  $144$

2. You are buying snacks. You buy 4 apples and a juice. The juice costs $1.75. The total cost is $4.75. How much is 1 apple?

   A  $.50

   B  $.75

   C  $1.00

   D  $1.25

**Constructed Response**

3. The sum of the angle measures of a triangle is 180°. Find the measure of each angle.

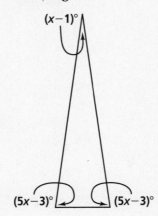

**Extended Response**

4. Write the correct explanation next to each step.

   • Distributive Property    • Combine like terms.    • Add 5 to each side and simplify.

   • Use multiplication to simplify.    • Multiply each side by $-\frac{5}{2}$ and simplify.

| Steps | Explanation |
|---|---|
| $-\frac{2}{5}(x+5) - 3 = 65$ | Original equation |
| $-\frac{2}{5}(x) - \frac{2}{5}(5) - 3 = 65$ | _____ |
| $-\frac{2}{5}(x) + (-2) - 3 = 65$ | _____ |
| $-\frac{2}{5}x - 5 = 65$ | _____ |
| $-\frac{2}{5}x = 70$ | _____ |
| $x = -175$ | |

# Common Core Standards Practice

## Week 4

### Selected Response

**1.** Solve the formula $pV = nRT$ for $n$.

**A** $n = pV - RT$

**B** $n = pVRT$

**C** $n = \dfrac{pV}{RT}$

**D** $n = pV$

### Constructed Response

**2.** You can find the net force $F$ on an object by using the formula $F = ma$, where $m$ is the mass of the object in kilograms and $a$ is its acceleration in meters per second squared. What is the mass of an object that has a net force of 65 kg m/s² and an acceleration of 5 m/s²?

### Extended Response

**3.** A T-shirt maker wants to open his first store. If he chooses the store on Main Street, he will pay $640 in rent and will charge $30 per t-shirt. If he chooses the store on Broad Street, he will pay $450 in rent and will charge $25 per T-shirt. How many T-shirts would he have to sell in 1 month to make the same profit at either location?

**a.** Write an equation to solve the problem.

**b.** Solve the equation you wrote in part (a) to answer the question.

# Common Core Standards Practice

## Week 5

**Selected Response**

**1.** Solve for *x*.

$$\frac{x}{4} = \frac{9}{12}$$

A  3

B  4

C  8

D  12

**2.** A 15-ft tree casts an 18-ft shadow at the same time that a 24-ft tree casts a shadow. How long is the shadow of the 24-ft tree?

A  20 ft

B  21.6 ft

C  27 ft

D  28.8 ft

**Constructed Response**

**3.** An architect builds a scale model of a skyscraper for a land development proposal. The model is 2 ft tall. The scale of the model is 1 in. : 12.3 m.

How tall is the proposed skyscraper in meters? Show your work.

**Extended Response**

**4.** A department store advertises a sale where the customer chooses the discount. A customer may choose a flat discount of $15 off any purchase or 15% off the total purchase price. The final purchase price was $150.

**a.** What are the possible amounts the customer spent without the discount?

**b.** Describe whether the customer chose the better discount. Justify your reasoning.

# Common Core Standards Practice

## Week 6

**Selected Response**

1. Solve $-3x \geq -15$.

   A $x \geq -5$
   B $x \geq 5$
   C $x \leq 5$
   D $x \geq -12$

2. Which inequality is graphed below?

   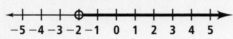

   A $b - 3 \geq 1$
   B $b + 3 < 1$
   C $b - 3 \leq 1$
   D $b + 3 > 1$

**Constructed Response**

3. Solve and graph the inequality $-3(g + 5) > 3$. Show your work and explain each step.

**Extended Response**

4. The length of a rectangle is 3 more than twice its width. If the perimeter of the rectangle can be no more than 78 ft, what are all of the possible widths of the rectangle?

   a. Write an inequality to solve the problem.

   b. Show your solution as a graph and describe the solution in words.

## Common Core Standards Practice                    Week 7

**Selected Response**

1. Graph the compound inequality $x > -3$ and $x \leq 5$.

A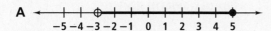

B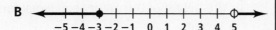

C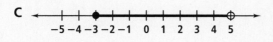

D

**Constructed Response**

2. Solve and graph $|x + 8| = 2$. Show your work.

**Extended Response**

3. John wants to raise between $200 and $300 for charity. His parents donated $50. John plans to ask others to contribute $10 each. How many people will need to contribute for John to reach his goal?

   **a.** Write an inequality to solve the problem.

   **b.** Show your solution as a graph and explain your solution in words.

# Common Core Standards Practice                    Week 8

**Selected Response**

1. Which equation represents the data in the table?

   **Paint Cans Remaining per Hour**

   | Number of Hours Worked, h | Number of Paint Cans Remaining, p |
   |---|---|
   | 0 | 10 |
   | 1 | 7 |
   | 2 | 4 |
   | 3 | 1 |

   **A** $p = 10h - 3$
   **B** $p = 3h - 10$
   **C** $p = 10 - 3h$
   **D** $p = 3 - 10h$

**Constructed Response**

2. The function rule $h = 3w + 2$ represents the height in centimeters $h$ of a plant after $w$ weeks of growth.

   **a.** Make a table of ordered pairs to show the height of the plant each week for 5 weeks.

   **b.** Graph the function. If the plant continues to grow at the same rate, how tall will it be after 8 weeks?

   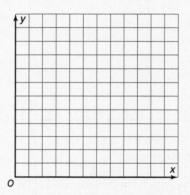

**Extended Response**

3. The posted rates for cab fare are $4 plus $1 per mile.

   **a.** Write an equation that represents total cab fare $c$ for $m$ miles.

   **b.** Sketch a graph of the equation on the axes provided.

   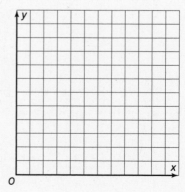

   **c.** How many miles can a passenger travel in the cab for $15?

# Common Core Standards Practice          Week 9

## Selected Response

1. Which of the following is NOT a function?

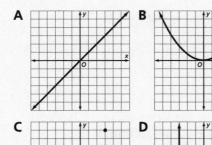

## Constructed Response

2. Consider the relation $\{(7, 10), (7, 17), (7, 24), (7, 31), (14, 10)\}$.

   a. What is the domain the of the relation?

   b. What is the range of the relation?

   c. Is the relation a function? Explain.

## Extended Response

3. A bowling alley charges $3 for the first game and $1.50 for each additional game.

   a. Write a function rule that represents this situation.

   b. How much does it cost to bowl 4 games? Show your work.

   c. How many games can be played for a total cost of $13.50? Show your work.

# Common Core Standards Practice                    Week 10

**Selected Response**

1. What is a recursive formula for the sequence 3, 24, 45, 67, 88, . . . ?

   A  $f(1) = 3; f(n) = f(n-1) + 21$
   B  $f(1) = 3; f(n) = f(n-1) - 21$
   C  $f(1) = 3; f(n) = f(n+1) + 21$
   D  $f(1) = 3; f(n) = f(n+1) - 21$

**Constructed Response**

2. The sequence $\frac{1}{2}, \frac{7}{8}, 1\frac{1}{4}, 1\frac{5}{8}, 2$ is an arithmetic sequence.

   a. Write an explicit function for the sequence.

   b. Find the value of the 12th term of the sequence.

**Extended Response**

3. Stacy opens a savings account with a deposit of $100. She deposits $25 each week.

   a. Write an explicit function and a recursive function to represent this situation.

   b. Choose a function from part (a) and determine how much money will be in the account after 15 weeks if Stacy makes no additional deposits or withdrawals.

   c. Explain why you chose the function you used in part (b).

Name _____ Class _____ Date _____

# Common Core Standards Practice                    Week 11

## Selected Response

1. Find the slope of the line that passes through the points $(-1, 6)$ and $(2, 15)$.

   A  $-9$            B  $-3$
   C  $3$             D  $9$

2. $y$ varies directly with $x$. $y = -6$ when $x = 2$. Write a direct variation equation that relates $y$ and $x$. Then find $y$ when $x = 4$.

   A  $y = 3x; y = 12$
   B  $y = -3x; y = -12$
   C  $y = 4x; y = 16$
   D  $y = -4x; y = -16$

## Constructed Response

3. Graph a line that has an $x$-intercept of 4 and a $y$-intercept of 3.

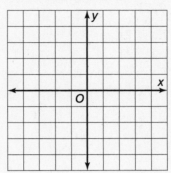

## Extended Response

4. A research team uses a generator to power some crucial items at its base camp. The researchers begin the expedition with 1200 gallons of gasoline for the generator. They plan to use 15 gallons of gas per day.

   a. Write an equation in slope-intercept form that relates the amount of gasoline $g$ remaining to the number of days $d$.

   b. Graph the equation from part (a).

   c. The team expects to use all of the gasoline during their expedition. How many days do they expect the expedition to last? How do you know?

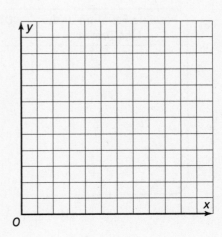

## Common Core Standards Practice

**Selected Response**

1. Which is an equation of a line in point-slope form that has slope 7 and passes through $(-2, 6)$?

   A  $y - 6 = 7(x + 2)$
   B  $y - 6 = 7(x - 2)$
   C  $y + 2 = 7(x - 6)$
   D  $y - 6 = -2(x - 7)$

**Constructed Response**

2. **a.** Draw a line on the graph below that has the same slope as the line drawn and that passes through $(0, -4)$.

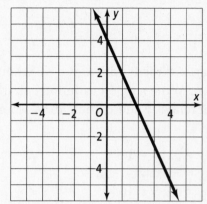

   **b.** What is an equation of the line you drew?

**Extended Response**

3. Refer to the graph at the right. Write each equation in the correct column.

   $y - 4 = \frac{1}{2}(x - 8)$        $y - 8 = \frac{1}{2}(x - 4)$

   $y - 1 = \frac{1}{2}(x + 10)$        $y + 10 = 2(x - 1)$

   $y + 10 = \frac{1}{2}(x - 1)$

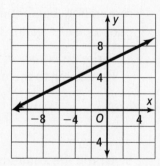

| Possible Equation of the Line Drawn | NOT an Equation of the Line drawn |
|---|---|
|  |  |

# Common Core Standards Practice          **Week 13**

## Selected Response

**1.** What is $y = -\frac{2}{3}x + 5$ written in standard form?

   **A** $3y = -2x + 5$

   **B** $3x - 2y = 15$

   **C** $2x + 3y = 15$

   **D** $2x - 3y = 5$

## Constructed Response

**2.** Are the graphs of $2x + 5y = 10$ and $5x - 6y = 6$ *parallel, perpendicular,* or *neither*? Justify your answer.

## Extended Response

**3.** Jesse and Lisa start a business tutoring students in math. They rent an office for $200 per month and charge $15 per hour per student.

   **a.** If they have 10 students each for one hour per week, how much profit do they make in a month? Write a linear equation to solve this problem.

   **b.** Graph the equation from part (a) and explain what it models.

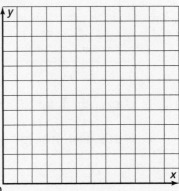

# Common Core Standards Practice

**Selected Response**

1. Which steps transform the graph of
   $y = |x|$ into the graph of $y = |x + 4| - 5$?

   **A** Translate 4 units right and 5 units up.
   **B** Translate 4 units left and 5 units up.
   **C** Translate 4 units right and 5 units down.
   **D** Translate 4 units left and 5 units down.

**Constructed Response**

2. **a.** Make a scatter plot and draw a trend line for the data at the right.

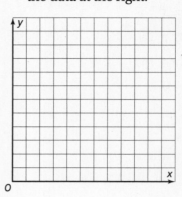

**Student Test Scores**

| Hours Spent Studying | Test Score |
|---|---|
| 3 | 70 |
| 6 | 88 |
| 2 | 68 |
| 7 | 90 |
| 1 | 60 |
| 4 | 73 |
| 8 | 92 |

   **b.** What would you expect a student who studied 5 hours to score on the test?

**Extended Response**

3. A teacher surveyed her students about the amount of physical activity they get each week. She then had their body mass index (BMI) measured.

   **a.** Use her data to make a scatter plot.

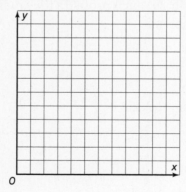

**Body Mass Index**

| Active Hours | BMI |
|---|---|
| 10 | 17 |
| 3 | 25 |
| 6 | 22 |
| 8 | 19 |
| 10 | 16 |
| 8 | 18 |
| 7 | 20 |

   **b.** Use a calculator to find the correlation coefficient.

   **c.** Is this relationship a correlation or causation or both? Explain how you know.

# Common Core Standards Practice

## Week 15

### Selected Response

**1.** What is the solution to the following system of equations?

$3y - 2x = 11$
$y + 2x = 9$

**A** $(2, 5)$
**B** $(5, 2)$
**C** $(-2, -5)$
**D** $(2, -5)$

### Constructed Response

**2.** Solve the following system of equations by using elimination. Show your work.

$x + 2y = 3$
$4x - 2y = 7$

### Extended Response

**3.** Antonio loves to go to the movies. He goes both at night and during the day. The cost of a matinee is $6. The cost of an evening show is $8. Antonio went to see a total of 5 movies and spent $36.

**a.** How many of each type of movie did he attend? Write a system of equations and solve by graphing.

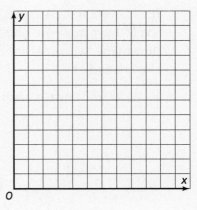

**b.** Why is the intersection of the graphs of the linear equations the solution?

# Common Core Standards Practice          Week 16

### Selected Response

1. Which ordered pair is NOT a solution of $y > 3x + 4$?

   A  (2, 12)
   B  (0, 5)
   C  (−2, 1)
   D  (1, 7)

### Constructed Response

2. Graph $3x + 2y > 6$.

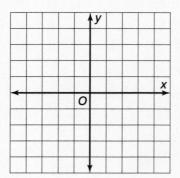

### Extended Response

3. The Movers scored a total of 80 points in their game last night against the Shakers. The Movers made no one-point shots, and a total of 35 two-point and three-point shots. How many two-point shots did the Movers make? How many three-point shots did the Movers make?

   a. Write and solve a system of equations to answer these questions.

   b. Graph the system of equations.

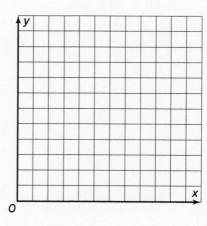

# Common Core Standards Practice                    Week 17

## Selected Response

**1.** Which is the graph of the solution for the system of inequalities?

$$y > -x + 3$$
$$y < x + 2$$

**A** 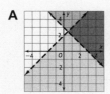     **B**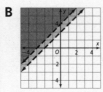

**C** 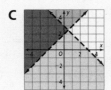     **D**

## Selected Response

**2.** Sketch the graph of a system of inequalities that has no solution. Describe how you know that the system has no solution.

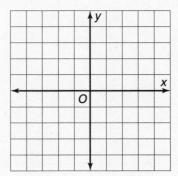

## Extended Response

**3.** Sarah is selling bracelets and necklaces to make money for her summer vacation. The bracelets cost $2 and the necklaces cost $3. She needs to make at least $500. Sarah knows that she will sell more than 50 bracelets.

**a.** Write a system of inequalities for Sarah's situation.

**b.** Graph the system of inequalities.

**c.** How many bracelets and necklaces could Sarah sell? Explain how you know that your solution is reasonable.

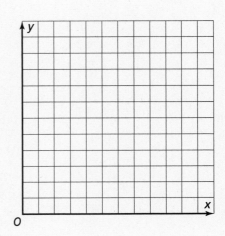

# Common Core Standards Practice                    Week 18

## Selected Response

**1.** What is the simplified form of $8^{-2}a^4 b^{-3}$?

   **A** $-16a^4 3b$

   **B** $-64a^4 b$

   **C** $\dfrac{a^4}{64b^3}$

   **D** $\dfrac{a^4}{16b^3}$

## Constructed Response

**2.** Simplify the expression

$\left(3a^{\frac{1}{5}} \cdot 4t^{\frac{2}{7}}\right)\left(2a^{\frac{4}{5}} \cdot t^{\frac{4}{7}}\right)$. Show your work.

## Extended Response

**3.** Your classmate writes that if $y > x$, then $\dfrac{a^x}{a^y} = \dfrac{1}{a^{(y-x)}}$ for all real numbers $x$ and $y$.

Is your classmate correct? Explain how you know and show examples to justify your explanation.

# Common Core Standards Practice         **Week 19**

## Selected Response

**1.** Write the expression $\sqrt[3]{8a^2} \cdot \sqrt[3]{8ab^5}$ in exponential form.

   **A** $8^{\frac{1}{3}}ab^2$

   **B** $8^{\frac{2}{3}}a^{\frac{1}{3}}b^{\frac{5}{3}}$

   **C** $4ab^{\frac{5}{3}}$

   **D** $64a^3b^5$

## Constructed Response

**2.** A population of prairie dogs doubles every year in the plains of North Dakota. The number of prairie dogs can be modeled by the equation $y = 400 \cdot 2^x$, where $x$ is the number of years after a scientist measures the population size. When $x = -5$, what does the value of $y$ represent?

## Extended Response

**3.** There are 8 mice in an attic. Their population is growing at a rate of 15% per month.

   **a.** Write an exponential growth equation to model this situation.

   **b.** How many mice will there be in the attic in two years if nothing is done to slow down or stop the growth?

   **c.** Sketch a graph of the function.

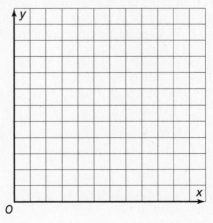

# Common Core Standards Practice

### Selected Response

1. Which is a recursive definition for the following geometric sequence?

$$3, 9, 27, 81,...$$

A $a_1 = 0; a_n = 3(a_{n-1})$

B $a_1 = 3; a_n = 3(a_{n-1})$

C $a_1 = 3; a_n = 3 + a_{n-1}$

D $a_1 = 0; a_n = 3 + a_{n-1}$

### Constructed Response

2. Brian has 2 parents, 4 grandparents, 8 great-grandparents, and so on.

   a. Write an explicit formula and recursive formula for the number of ancestors Brian has in a generation if he goes back to the $n$th generation.

   b. Interpret the parts of the formula and explain their meaning within the context of this situation.

### Extended Response

3. Suppose that a new house is worth $200,000 and that it depreciates at a rate of 15% a year.

   a. Explain this situation in terms of growth or decay.

   b. Construct a function to model this situation.

   c. Estimate the value of the house after 5 years.

# Common Core Standards Practice    Week 21

## Selected Response

1. Which expression is equivalent to $(x + 4)^2$ ?

   **A** $2x + 8$
   **B** $x^2 + 8x + 16$
   **C** $x^2 + 16$
   **D** $x^2 + 16x + 8$

## Constructed Response

2. Classify each expression as equivalent to $2x^2 + 17x$ or **NOT** equivalent to $2x^2 + 17x$. Write each expression inside the appropriate box below.

   $(4x^2 + 10x + 7) - (2x^2 - 7x + 7)$

   $x(2x + 17)$

   $5x^2 - (3x^2 + 12x) + 5x$

   $(2x^2 + 5x^2) + (-5x^2 + x + x + 15x)$

   | Expressions Equivalent to $2x^2 + 17x$ | Expressions Not Equivalent to $2x^2 + 17x$ |
   |---|---|
   |  |  |

## Extended Response

3. The length of a rectangular sandbox is $3x + 5$. The width of the sandbox is $x - 3$.

   **a.** What polynomial in standard form represents the area of the sandbox?

   **b.** Name the polynomial based on its degree and number of terms.

Name _____ Class _____ Date _____

# Common Core Standards Practice                    Week 22

**Selected Response**

1. What is the degree of the monomial $3x^4yz$?

   A 4

   B 6

   C 7

   D 8

**Constructed Response**

2. **a.** Simplify the expression $(x + 2)^3$.

   **b.** Name the polynomial based on its degree and number of terms.

**Extended Response**

3. Refer to the figure shown. When necessary, use 3.14 for $\pi$.

   **a.** What is an expression for the area of the rectangle? Simplify your answer.

   **b.** What is an expression for the area of the circle? Simplify your answer.

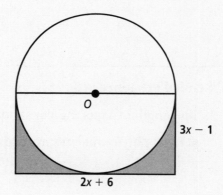

   **c.** What is an expression for the area of the shaded region? Simplify your answer.

   **d.** If the radius of the circle is 5 cm, what is the area of the shaded region?

# Common Core Standards Practice          **Week 23**

### Selected Response

1. Factor the following polynomial.

   $7x^4 - 4x^3 + 28x^2 - 16x$

   **A** $x(7x^2 - 4)(x + 4)$
   **B** $x(7x + 4)(x^2 - 4)$
   **C** $x(7x - 4)(x^2 + 4)$
   **D** $x(7x^2 + 4)(x - 4)$

### Constructed Response

2. The area of a square window is $36x^2 + 96x + 64$. What is a side length of the window?

### Extended Response

3. A rectangular garden measuring 4 m by 6 m is to have a pathway $x$ meters wide installed around its perimeter. The area of the pathway will be equal to the area of the garden.

   **a.** Make a sketch of this situation and define each part of the polynomial.

   **b.** What will be the width of the pathway?

# Common Core Standards Practice

# Week 24

## Selected Response

1. What is the vertex of the parabola with equation $y = 3x^2 + 1$?

   A  $(1, 3)$

   B  $(3, 1)$

   C  $(1, 0)$

   D  $(0, 1)$

## Constructed Response

2. **a.** Graph the equation $y = 4x^2 - 4$.

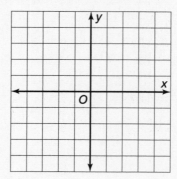

   **b.** Find the solutions to $4x^2 - 4 = 0$.

## Extended Response

3. You throw a ball into the air from a building. The ball's height $h$, in feet, after $t$ seconds can be modeled by the function $h(t) = -16t^2 + 32t + 48$.

   **a.** After how many seconds will the ball hit the ground? Solve by factoring and by sketching a graph.

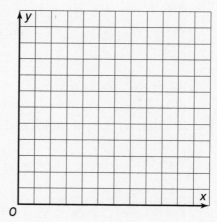

   **b.** Interpret the key features of the graph and how they relate to this situation.

# Common Core Standards Practice                     Week 25

### Selected Response

1. Choose all the solutions of $x^2 - 8 = 2x$.

   A  $x = -4$

   B  $x = -2$

   C  $x = 2$

   D  $x = 4$

### Constructed Response

2. **a.** Solve $x^2 + 4x = -1$ by completing the square.

   **b.** Use the zeros you found in part (a) to graph the function defined by the polynomial equation $x^2 + 4x = y - 1$.

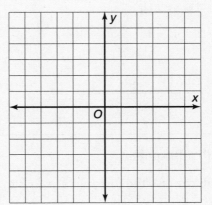

### Extended Response

3. The area of a rooftop can be expressed as $4x^2 + 4x + 1$. The rooftop is a quadrilateral.

   **a.** What expression describes the length of one side of the rooftop?

   **b.** What type of quadrilateral is the rooftop? How do you know?

   **c.** If the area of the rooftop is 441 m², what is $x$?

# Common Core Standards Practice

## Week 26

### Selected Response

**1.** Solve $x^2 + 20x = -40$ by completing the square.

**A** $x = -10 \pm 2\sqrt{15}$

**B** $x = 10 \pm 2\sqrt{15}$

**C** $x = -10 \pm 2\sqrt{35}$

**D** $x = 10 \pm 2\sqrt{35}$

**2.** Solve $x^2 + 4x - 2 = 3$ by using the quadratic formula.

**A** $x = 1, -5$

**B** $x = -1, 5$

**C** $x = 2, -10$

**D** $x = 4, -8$

### Constructed Response

**3.** The flight path of an eagle is modeled by the function $y = x^2 - 14x + 16$, where $y$ is the eagle's height in feet above the water and $x$ is time in seconds.

**a.** Solve the equation by completing the square. Show your work.

**b.** What does this solution tell you about the eagle's flight?

### Extended Response

**4.** You can use the formula $V = lwh$ to find the volume of a box.

**a.** Write a quadratic equation in standard form that represents the volume of the box.

**b.** The volume of the box is 6 ft$^3$. Solve the quadratic equation for $x$.

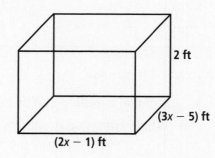

2 ft

$(3x - 5)$ ft

$(2x - 1)$ ft

**c.** Use the solution from part (b) to find the length and width of the box. Describe any extraneous solutions.

# Common Core Standards Practice

## Week 27

### Selected Response

**1.** What is the value of $x$?

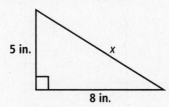

**A** $\sqrt{26}$ in.

**B** $2\sqrt{13}$ in.

**C** $\sqrt{74}$ in.

**D** $\sqrt{89}$ in.

**2.** What is the simplified form of $8\sqrt{12} + 3\sqrt{3}$ ?

**A** $19\sqrt{3}$

**B** $11\sqrt{15}$

**C** $19\sqrt{6}$

**D** $35\sqrt{3}$

### Constructed Response

**3.** What are the solutions of $\sqrt{9x + 1} = x + 2$ ?

### Extended Response

**4.** The length of side $d$ can be expressed as $d = \sqrt{3^2 + a^2}$.

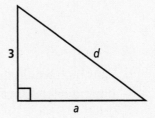

**a.** Write an equation that represents the length of side $a$ in terms of $d$.

**b.** If $d = 5$, what are all possible values of $a$?

**c.** Are any of the values you found in part (b) extraneous solutions? Explain.

# Common Core Standards Practice     Week 28

## Selected Response

1. Which steps transform the graph of $y = \sqrt{x}$ to $y = 2\sqrt{x + 3} - 4$? Select all that apply.

   **A** Stretch vertically by the factor 2.

   **B** Stretch vertically by the factor $\frac{1}{2}$.

   **C** Translate 3 units to the right.

   **D** Translate 3 units to the left.

   **E** Translate 4 units up.

   **F** Translate 4 units down.

## Constructed Response

2. **a.** Graph the function $y = \sqrt{x + 5} - 3$.

   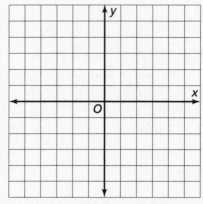

   **b.** What are the domain and range of the function?

## Extended Response

3. The distance a person can see through a particular submarine periscope is given by the equation $d = 6\sqrt{h - 4} + 3$, where $h$ is the height in feet above water.

   **a.** Graph the equation.

   **b.** How high would the submarine periscope have to be to spot a ship 27 mi away?

   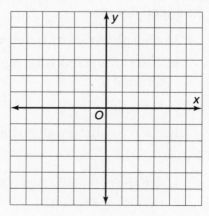

# Common Core Standards Practice

## Week 29

**Selected Response**

1. The length of a rectangle is $3x^2 + 5x + 2$ units and the area is $12x^3 + 23x^2 + 13x + 2$ square units. Write an expression for the width of the rectangle.

   **A** $4x - 1$ units

   **B** $x + 4$ units

   **C** $4x + 1$ units

   **D** $x - 4$ units

**Constructed Response**

2. Classify the following equations for those that have $y = 4$ or $x = 6$ as asymptotes of their graphs and for those that have neither as asymptotes. Write each equation in the appropriate column in the table below. Some may belong in more than one column.

$$y - 4 = \frac{1}{x - 6} \qquad y = \frac{4}{x - 6}$$

$$4y = \frac{1}{x + 6} \qquad y = \frac{(x + 2)}{(x + 2)(x - 3)} + 4$$

| Graphs With $y = 4$ as Asymptote | Graphs With $x = 6$ as Asymptote | Graphs With Neither $y = 4$ nor $x = 6$ as Asymptotes |
|---|---|---|
|  |  |  |

**Extended Response**

3. Andrew and Bill, working together, can cover the roof of a house in 6 days. Andrew, working alone, can complete the job in 5 days less than Bill. How long will it take Bill to complete the job?

   **a.** Write and solve a quadratic function to model this situation.

   **b.** Interpret your results in the context of this situation.

# Common Core Standards Practice         Week 30

## Selected Response

1. Which histogram is uniform?

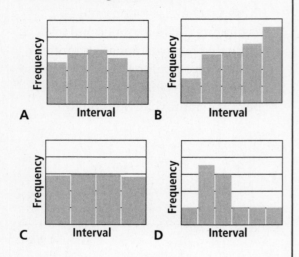

A        B

C        D

## Constructed Response

2. The hours that a school band practiced per week are listed below.

   7 5 9 7 4 6 10 8 5 7 8 7 3 12 15 13 8

   a. What are the mean, median, mode, and range of their practice times?

   b. Which measure of central tendency best describes their practice times? Justify your answer.

## Extended Response

3. The table below shows a company's automobile sales for the first two quarters of the year.

| Quarter | Mini Buggy | Overhaul 4 × 4 |
|---------|------------|----------------|
| 1 | 108 | 216 |
| 2 | 198 | 140 |

Calculate the ratio and percent for each of the following situations.

a. Mini Buggy sales in quarter 1 to all Mini Buggy sales

b. Mini Buggy sales in quarter 1 to Mini Buggy sales in quarter 2

c. Overhaul 4 × 4 sales in quarter 2 to all Overhaul 4 × 4 sales

d. Overhaul 4 × 4 sales in quarter 1 to all automobile sales in both quarters

e. Which automobile's sales were highest in quarter 1? In quarter 2?

# End-of-Course Assessment

## Selected Response

**Read each question. Then circle the letter(s) of the correct answer(s).**

1. A student is 5 ft 9 in. tall. Which of the following are equivalent to the student's height? Use the fact that 1 m ≈ 3.28 ft.

   **A** 1.75 m

   **B** 17.5 cm

   **C** 1750 cm

   **D** 1750 mm

2. Which is the simplified form of the expression?

   $9(r + 3) - \frac{1}{2}(4r - 16)$

   **A** $11r + 19$

   **B** $11r + 35$

   **C** $7r + 19$

   **D** $7r + 35$

3. Which model is most appropriate for the set?

   $(2, 17), (6, 121), (0, 1), (3, 34), (-1, 2), (-7, 0)$

   **A** Linear

   **B** Exponential

   **C** Quadratic

   **D** Logarithmic

4. If $f(x) = \frac{3}{4}x + \frac{5}{6}$, what is $f(12)$?

   **A** $\frac{59}{6}$

   **B** $\frac{47}{6}$

   **C** $\frac{23}{6}$

   **D** 19

5. Which function rule is graphed below?

   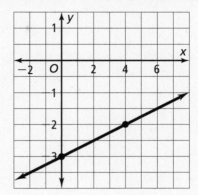

   **A** $y = 3x - 4$

   **B** $y = \frac{1}{4}x - 3$

   **C** $y = -\frac{1}{4}x - 3$

   **D** $y = \frac{1}{2}x - 3$

6. Solve $y = 2xz^2 - xy$ for $x$.

   **A** $x = \frac{1}{2z^2}$

   **B** $x = \frac{y}{2z^2}$

   **C** $x = \frac{1}{2z^2 - 1}$

   **D** $x = \frac{y}{2z^2 - y}$

7. Which of the following are equivalent to the polynomial?

   $4x(2x^2 - 1) + x(8x)$

   **A** $4x(2x^2 + 2x - 1)$

   **B** $8x^3 + 8x^2 - 4x$

   **C** $6x^3 + 8x^2 - 4x$

   **D** $8x^3 - 8x^2 + 4x$

**8.** What is the factored form of
$3x^2 - 17x = 28$?

   **A** $(x + 7)(3x - 4) = 0$

   **B** $(x + 7)(3x + 4) = 0$

   **C** $(x - 7)(3x - 4) = 0$

   **D** $(x - 7)(3x + 4) = 0$

**9.** Which of the following give a definition for the geometric sequence?
$3, -3, 3, -3, 3, \ldots$

   **A** $a_1 = -3; a_n = a_{n-1} \cdot -1$

   **B** $a_n = 3 \cdot (-1)^{n-1}$

   **C** $a_1 = 3; a_n = a_{n-1} \cdot 1$

   **D** $a_1 = 3; a_n = a_{n-1} \cdot -1$

**10.** Mandy works part-time to earn money for a trip. The amount she saves after working $x$ hours is given by the equation $y = 7.5x + 40$. How much does Mandy earn per hour?

   **A** $7.50

   **B** $32.50

   **C** $40

   **D** $47.50

**11.** Express the following sentence in equation form.

Five times the difference of a number and 2 is equal to the quotient of the same number and 6.

   **A** $5x - 2 = \frac{x}{6}$

   **B** $5(x - 2) = \frac{x}{6}$

   **C** $5(2 - x) = \frac{x}{6}$

   **D** $5(x - 2) = \frac{6}{x}$

**12.** Which of the following result in an irrational number?

   **A** the sum of two rational numbers

   **B** the product of two rational numbers

   **C** the sum of a rational number and an irrational number

   **D** the product of a nonzero rational number and an irrational number

**13.** The graph of which equation is shown below?

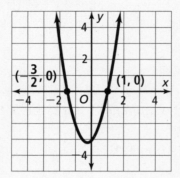

   **A** $y = 2x^2 + x - 3$

   **B** $y = \frac{1}{2}x^2 + x - 3$

   **C** $y = 4x^2 + x - 3$

   **D** $y = x^2 + x - 3$

**14.** The graph of which equation is shown below?

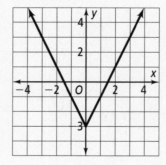

   **A** $y = -|2x| + 3$

   **B** $y = |2x| - 3$

   **C** $y = |2x - 3|$

   **D** $y = 2|x - 3|$

**15.** Solve the equation $2x^2 + 3x = 2$.

    **A** $-2$

    **B** $-\frac{1}{2}$

    **C** $\frac{1}{2}$

    **D** $2$

**16.** Which of the following are equivalent forms of the equation $y = -\frac{5}{9}x + \frac{2}{3}$?

    **A** $y + 1 = -\frac{5}{9}(x - 3)$

    **B** $5x + 9y = 6$

    **C** $\frac{5}{9}x + y = \frac{2}{3}$

    **D** $-5x + 9y = 6$

**17.** An engineer studied the sales of trucks and SUVs in California over an 8-year period. The results are modeled in thousands sold with the following polynomials.

Trucks: $-13x^3 + 89x^2 - 119x + 6814$
SUVs: $6x^2 - 12x + 2152$

In each polynomial, $x = 0$ corresponds to the first year in the 8-year period. Which polynomial models the total number of trucks and SUVs sold in California during the 8-year period?

    **A** $-13x^3 + 105x^2 - 131x + 8966$
    **B** $-13x^3 + 73x^2 - 107x + 4662$
    **C** $-13x^3 + 105x^2 - 107x + 8966$
    **D** $13x^3 - 73x^2 + 107x - 4662$

**18.** At which point do the graphs of the equations intersect?

$$\begin{cases} y = 3x - 5 \\ y = |x - 7| \end{cases}$$

    **A** $(-1, 8)$

    **B** $(3, 4)$

    **C** $(-1, -8)$

    **D** $(3, -4)$

**19.** Which of the following correlations represents a causal relationship?

    **A** the number of cats and the number of dogs in a shelter

    **B** the number of cats in a shelter and the amount of cat food used

    **C** the number of cats in a shelter and the number of vaccinations given

    **D** the amount of money in the cash drawer and the number of cats in the shelter

**20.** Which of the following are solutions to the inequality $-9 \le 2x + 1 \le 5$

    **A** $-6$         **D** $0$

    **B** $-4$         **E** $2$

    **C** $-2$         **F** $4$

**21.** Which of the following are solutions to the inequality $x^2 - 4 = x + 8$

    **A** $-4$

    **B** $-3$

    **C** $3$

    **D** $4$

**22.** Which of the following are equivalent to 18 feet per minute?

    **A** 3.6 inches per hour

    **B** 40 inches per minute

    **C** 1080 feet per hour

    **D** 12,960 inches per hour

**23.** Which points are in the solution set for $4x - y > 1$?

    **A** $(1, 2)$      **D** $(2, 0)$

    **B** $(0, -1)$      **E** $(-1, 2)$

    **C** $(0, 2)$      **F** $(-1, -2)$

**24.** At which points do the graphs of the following equations intersect?

$$\begin{cases} y = x^2 + 9x + 1 \\ x - y = 6 \end{cases}$$

    **A** $(0, -6)$      **E** $(-1, -7)$

    **B** $(0, 1)$      **F** $(-1, 0)$

    **C** $(-4, -9)$      **G** $(-7, -13)$

    **D** $(-4, -19)$      **H** $(-7, 0)$

**25.** What are the factors of the expression $x^2 - 7x - 44$?

    **A** $(x - 11)$

    **B** $(x - 4)$

    **C** $(x + 4)$

    **D** $(x + 11)$

## Constructed Response

**In this section, show all your work in the space beneath each test item.**

**26.** Solve the equation. Show your work and justify each step.

$$4\left(x - \frac{1}{2}\right) - 12 = 0$$

**27.** The perimeter of a rectangle is $6x^2 - 6x - 4$. The width of the rectangle is $2x + 1$. What is the length of the rectangle?

**28.** Graph the inequality $4x + 2y < 6$.

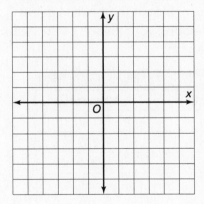

**29.** Graph the function $f(x) = -x^2 + 2x + 4$.

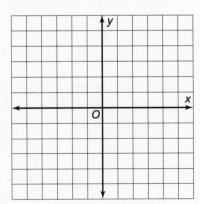

**30.** The ages of the members of a hiking club are 17, 18, 24, 28, 32, 36, 43, and 52. A new members who is 45 years old joins the club. In general, describe how this will affect the mean, median, mode, and range of the ages of the members of the club.

**31.** What are the solutions of the equation $x^2 - 4x - 32 = 0$? Show your work and justify each step.

**32.** One angle of an obtuse triangle measures four times the first angle. The third angle measures 30° less than the first angle. What are the degree measures of the three angles? Show or explain your work.

**33.** What is the solution to the system of equations?

$$\begin{cases} y = 2x + 3 \\ y = -x + 6 \end{cases}$$

**34.** You buy $x$ pounds of strawberries for \$3.99/lb. Write a function rule for the amount of change $C$ you receive from a \$20 bill.

**35.** Write an equation for the line of best fit for the scatter plot below.

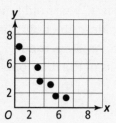

**36.** A square has sides measuring $5\sqrt{9}$ m. What is the area of the square?

**37.** Suppose you survey each coach at a cheerleading tournament. What relationship would you expect between the number of coaches and the number of teams competing in the tournament?

**38.** The population of a town is 75,000 and decreases 1.5% each year. If the trend continues, what will the population be after 12 years? Round your answer to the nearest thousand.

**39.** A beach club made $39,100 in May and $59,200 in August. What is the rate of change in the profit for this time period?

**40.** Write a sequence that is both arithmetic and geometric.

**41.** What function does the table represent?

| $x$ | $-2$ | $-1$ | 0 | 1 | 2 |
|---|---|---|---|---|---|
| $y$ | 4 | 5 | 6 | 7 | 8 |

**42.** What is the value of the function $f(x) = \frac{1}{3}(-5x) + 3$ when $x = 0.25$?

**43.** What is the correlation coefficient of the line of best fit for the data in the table? Round your answer to the nearest thousandth.

| Attendance at Water Park | |
|---|---|
| **Month** | **Attendance** |
| April | 130 |
| May | 276 |
| June | 874 |
| July | 951 |
| August | 712 |
| September | 402 |

**44.** A puddle is 0.06 m deep after 1 h and 0.03 m deep after 5 h. At what rate is the level of the water changing?

**45.** How does the graph of $y = 3x - 1$ differ from that of $y = 3x$?

**46.** Write an inequality for the graph below.

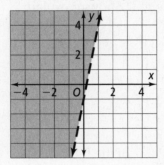

**47.** What is the solution of the system of equations?

$$\begin{cases} y = 4x - 1 \\ y = 3x + 2 \end{cases}$$

**48.** What is the mean of the data?
$6x, 3x, 17x, 4x, 10x, 2x$

**49.** What is the standard deviation of the data set rounded to the nearest thousandth?

7.2, 9.1, 5.7, 8.5, 10.2, 9.9, 11.0, 7.7, 6.4, 8.9

**50.** What is the vertex of the graph of the function $f(x) = x^2 + 4x - 5$?

**51.** What is the value of $x$? Explain each step in your solution.

$$\tfrac{1}{10}(1.2x - 3.5) = 0.13$$

**52.** Is the relation a function? Explain how you know.

{(300, 9), (260, 4), (275, 4), (350, 11), (225, 2), (300, 7), (325, 10), (280, 5)}

**53.** Write an explicit formula for the arithmetic sequence.
$\tfrac{2}{3}, \tfrac{3}{4}, \tfrac{5}{6}, \tfrac{11}{12}, 1, \ldots$

**54.** In the following situation, is there likely to be a correlation? If so, does the situation reflect a causal relationship?

the cost per pound of salad at a salad bar and the amount of salad sold

**55.** In May, your savings account balance was $1140. In August, the balance in the account was $1450. What is the average rate of change per month?

## Extended Response

**In this section, show all your work in the space beneath each test item.**

56. A ball is thrown directly upward from a height of 30 ft with an initial velocity of 64 ft/s. The equation $h = -16t^2 + 64t + 30$ gives the height $h$ after $t$ seconds.

   **a.** How long does it take for the ball to reach its maximum height? Show or explain your work.

   **b.** What values can be used for the domain?

57. Refer to the table below.

| x | 1 | 1.5 | 2 | 2.5 | 3 | 3.5 | 4 |
|---|---|-----|---|-----|---|-----|---|
| y | 5 | 6 | 5 | 7 | 6.5 | 7.5 | 8 |

   **a.** Make a scatter plot of the data.

   **b.** Estimate an equation of the line of best fit.

**58.** A banquet hall charges $750 to feed large parties. For a family reunion, the cost will be divided equally among each attending family member. Each person also must pay $3.50 for a tip.

    **a.** Approximately how many people must attend the reunion in order for the total cost per person to be about $15 per person?

    **b.** Describe the change in the cost per person as the number of family members who attend the reunion increases.

**59.** The table shows the average weight for girls between 2 and 7 years old.

| Average Weight for Girls | | | | | | |
|---|---|---|---|---|---|---|
| Age (years) | 2 | 3 | 4 | 5 | 6 | 7 |
| Weight (lb) | 28.4 | 30.8 | 35.2 | 39.6 | 46.2 | 50.6 |

    **a.** What is the slope of a trend line for the data rounded to the nearest tenth? What does the slope tell you about the situation in this problem?

    **b.** What is the correlation coefficient of age and weight to three decimal places? What does the correlation coefficient tell you about the situation in this problem?

**60.** Consider the equation $x^2 + 4x + 1$.

    a. What are the solutions to the equation?

    b. Explain why completing the square is a better strategy for solving this equation than graphing or factoring.

**61.** A farmer records his profits for the week from selling corn at the farmers market.

| Corn Profits | | | | |
|---|---|---|---|---|
| Corn sold (lb) | 22 | 31 | 66 | 73 |
| Profit ($) | 132 | 186 | 396 | 438 |

    **a.** Do the values in the table represent a function? Explain how you know.

    **b.** How much profit does the farmer make for each pound of corn? How many pounds of corn would the farmer need to sell to earn a profit of $540?

**62.** A student found the solutions of $0 = x^2 - 5x - 24$ to be $x = 3$ and $x = -8$.

    **a.** Is the student correct? If not, what are the solutions of the equation?

    **b.** Graph the equation $y = x^2 - 5x - 24$ on the grid below.

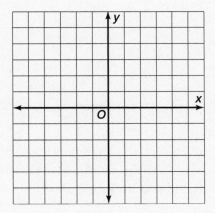

**63.** A cell phone plan costs $35 per month plus 5 cents for each minute of use.

    **a.** Write a function for the cost of the plan. What are the domain and range of the function?

    **b.** How much would you expect your monthly bill to be if you used 345.6 minutes last month?

**64.** The surface area of a cylinder is given by the equation $2\pi r^2 + 2\pi rh$, where $r$ is the radius of the cylinder and $h$ is the height of the cylinder.

   **a.** What is the surface area of a cylinder with radius $x - 1$ and height $2x$?

   **b.** Find the surface area of the cylinder from part (a) if $x = 11$.

**65.** Consider this system of equations.

$$\begin{cases} 7x + 2y = 16 \\ -21x - 6y = 24 \end{cases}$$

   **a.** Find the solution of the system of equations using substitution.

   **b.** Solve both equations for $y$. What can you say about the graphs of these equations?

# Performance Task: Choosing a Movie-Rental Plan

**Complete this performance task in the space provided. Fully answer all parts of the performance task with detailed responses. You should provide sound mathematical reasoning to support your work.**

You are considering three different ways to rent movies.

**Plan A:** Rent DVDs from a kiosk in a nearby grocery store for $1.50 each. The selection of movies is limited.

**Plan B:** Stream unlimited movies to your computer or TV for $10 per month. The selection of movies is good.

**Plan C:** Rent DVDs by mail for a $5 monthly fee plus $2 per movie. The selection of movies is outstanding.

## Task Description

Choose the movie-rental plan that you think is best. Consider the cost of each plan, the selection offered, and how you like to receive and watch movies.

**a.** Write functions $A(x)$, $B(x)$, and $C(x)$ that give the cost to rent $x$ movies per month for Plans A, B, and C, respectively.

# Performance Task: Choosing a Movie-Rental Plan (continued)

**b.** If you consider only cost, under what condition does it make sense to choose Plan B over Plan A?

**c.** If you consider only cost, under what condition does it make sense to choose Plan C over Plan B?

**d.** Show that Plan A is always more cost-effective than Plan C. Does that mean that Plan A is a better choice than Plan C for everyone? Explain.

**e.** Which movie-rental plan would you choose? Justify your answer.

# Performance Task: Expanding a Parking Lot

**Complete this performance task in the space provided. Fully answer all parts of the performance task with detailed responses. You should provide sound mathematical reasoning to support your work.**

A high school has a rectangular parking lot that measures 600 ft long by 400 ft wide. The school board wants to double the area of the lot by increasing both its length and width by the same amount, $x$ ft. The board also wants to build a fence around the new lot. The cost to expand the lot is estimated to be $2 per square foot of new space. The cost to fence the lot is estimated to be $30 per foot of fencing. Costs include materials and labor.

## Task Description

Estimate the total cost of expanding and fencing in the lot.

**a.** Draw a diagram of the situation. Your diagram should show both the original parking lot and what the lot will look like after it has been expanded. Label all dimensions.

**b.** Write an equation that you can use to find $x$. Solve the equation for $x$.

# Performance Task: Expanding a Parking Lot (continued)

**c.** What is the area of the new portion of the parking lot that needs to be built? What is the perimeter of the new parking lot?

**d.** What is the estimated cost of expanding and fencing in the lot?

**e.** The school has only enough money to pay for half the estimated cost from part (d). The school board plans to raise the remaining funds by selling parking stickers for $100 to students and $200 to faculty. How many student stickers and how many faculty stickers must the school sell? Is there only one possible answer? Explain.

# Performance Task: Projectile Motion

**Complete this performance task in the space provided. Fully answer all parts of the performance task with detailed responses. You should provide sound mathematical reasoning to support your work.**

Suppose an object is launched at an angle of 45° with respect to horizontal. The object's height $y$ (in feet) after it has traveled a horizontal distance of $x$ feet is given by the equation

$$y = -\frac{g}{v^2}x^2 + x + y_0$$

where $v$ is the object's initial speed (in feet per second), $y_0$ is the object's initial height (in feet), and $g \approx 32$ ft/s$^2$ is the acceleration due to gravity.

## Task Description

You throw a baseball at a 45° angle to your friend standing 100 ft away. Your friend holds her glove 5 ft above the ground to catch the ball. At what initial height, and with what initial speed, should you release the ball so that your friend can catch it without moving her glove?

**a.** Suppose you release the baseball with an initial height of 6 ft and an initial speed of 50 ft/s. Write and graph an equation that represents the ball's path. Does the ball land in your friend's glove? Explain.

# Performance Task: Projectile Motion (continued)

**b.** If your friend catches the ball, what point must lie on the graph of the ball's path? (Assume you are standing at the point (0, 0).)

**c.** Use your answer to part (b) to write an equation that describes the initial heights and initial speeds for which your friend catches the ball. Explain why there is more than one initial height and initial speed that work.

**d.** Find an initial height and an initial speed for which your friend catches the ball. How can you check your answer?

# Performance Task: Calculating Inflation

Complete this performance task in the space provided. Fully answer all parts of the performance task with detailed responses. You should provide sound mathematical reasoning to support your work.

The inflation rate for an item (such as a carton of eggs) measures how rapidly the price of the item has changed over time. If an item's price changes from $p_1$ to $p_2$ over a period of $n$ years, then the annual inflation rate $r$ (expressed as a decimal) is given by this equation:

$$r = \left(\frac{p_2}{p_1}\right)^{\frac{1}{n}} - 1$$

For example, if the price of an item increases from $2 to $3 over 5 years, then the annual inflation rate is:

$$r = \left(\frac{3}{5}\right)^{\frac{1}{5}} - 1 \approx 1.084 - 1 = 0.084 = 8.4\%$$

The table below shows the average retail prices of several foods in the United States for the years 2000 and 2009.

| Food | Price in 2000 | Price in 2009 |
|------|---------------|---------------|
| Bread (1 lb) | $0.99 | $1.39 |
| Butter (1 lb) | $2.80 | $2.67 |
| Cheddar cheese (1 lb) | $3.76 | $4.55 |
| Eggs, large (1 dozen) | $0.96 | $1.77 |
| Ground beef (1 lb) | $1.63 | $2.19 |
| Oranges (1 lb) | $0.62 | $0.93 |
| Peanut butter (1 lb) | $1.96 | $2.10 |
| Meat cutlets (1 lb) | $3.46 | $3.29 |
| Tomatoes (1 lb) | $1.57 | $1.96 |

## Task Description

Identify the foods in the table with the least and greatest annual rates of inflation for the period 2000–2009. Then predict the cost in 2015 of the groceries in a basket containing one of each food item from the table.

Name _____ Class _____ Date _____

# Performance Task: Calculating Inflation (continued)

**a.** Find the annual inflation rate for each food in the table for the period 2000–2009. Round your answers to the nearest tenth of a percent.

**b.** Which food had the least inflation rate? Which had the greatest inflation rate?

**c.** Which foods, if any, had a negative annual inflation rate? What does a negative annual inflation rate mean?

**d.** Can you add the percentages of each item to determine the inflation rate for a basket of groceries? Explain.

**e.** What was the annual inflation rate for a basket of all the food items for the period 2000–2009?

**f.** Predict the cost of the groceries in the basket in 2015. Explain how you determined your prediction.